Confronting Disaster

Confronting Disaster

An Existential Approach to Technoscience

Raphael Sassower

LEXINGTON BOOKS
Lanham • Boulder • New York • Toronto • Oxford

LEXINGTON BOOKS

Published in the United States of America
by Lexington Books
An imprint of The Rowman & Littlefield Publishing Group, Inc.
4501 Forbes Boulevard, Suite 200, Lanham, Maryland 20706

PO Box 317
Oxford
OX2 9RU, UK

British Library Cataloguing in Publication Information Available

Library of Congress Cataloging-in-Publication Data

Sassower, Raphael.
Confronting disaster : an existential approach to technoscience / Raphael Sassower.
p. cm.
Includes bibliographical references and index.
ISBN 0-7391-0850-6 (cloth : alk. paper)—ISBN 0-7391-0851-4 (pbk. : alk. paper)
1. Technology—Social aspects. 2. Technology and civilization. 3. Science—Social aspects. 4. Science and civilization. I. Title.
T14.5.S26 2004
303.48'3—dc22 2004000458

Printed in the United States of America

∞™ The paper used in this publication meets the minimum requirements of American National Standard for Information Sciences—Permanence of Paper for Printed Library Materials, ANSI/NISO Z39.48–1992.

~

Contents

Acknowledgments

I wish to thank my colleague Jeffrey Rubin-Dorsky, writer Katie Johnson, my friend Bonnie Lessing, and editor Maura Allaby, who have helped focus my writing away from esoteric ideas and onto what might be an important message in this book. I'm indebted to them for their thoughtful comments. I'm grateful to my academic institution for, as usual, staying out of my intellectual way.

CHAPTER ONE

~

Introduction

Science and technology have permeated every aspect of our daily lives. The basic ambiguities these scientific discoveries have unearthed (for example, the subatomic structure of the universe and the genetic composition of the human body) coupled with the anxieties of the two world wars have brought us more anguish than peace at the end of the previous century. Just think of your use of cars, computers, television sets, cellular telephones, credit cards, and the Internet, as well as your encounters with health care providers (nurses, doctors, clinics, vitamins, prescription drugs, and even insurance companies that manage our health). Given the influence of science and technology, the traditional appeal to church or state seems less reassuring than it was in relation to anything from, say, the pollution of water supply or the potential for success in applying battlefield strategies. Do we ask our priest or rabbi what to do about our medical condition? Do we expect the president to deal with our company's computerized surveillance of our lives? Any promise by politicians for quick relief is met with disbelief or outright incredulity, even when the relief is economic in nature. Our mistrust runs so deep that we revert back to family and friends for reassurance.

What each of us seeks more than anything else—psychologists, sociologists, and anthropologists will agree—is some sense of personal stability and security. Will the church take care of us when everything is said and done, or will it only care about how much we contribute this week? Will our country take care of us when we are sick and unemployed or when we retire? Even when the policies are in place, can we count on them to be implemented in a way that will help our own unique situation? At times we measure our security in financial terms, at times in terms of personal relationships. We have enough money saved, so we can take care of ourselves. But is this indeed enough money after all? Our children love us and care about our health, but will they invite us to their homes when we get old and disabled? We know how tentative our security is and can at least fall back on some kind of order in our immediate environment. So, perhaps the word is not security but order. Order (as a coping method) in this sense is the way our lives and our environment are organized by us. The way we reorganize the pieces of the puzzle is the way we put this puzzle together so as to have a full picture of our lives. The pieces range from our interactions with other people (including family members) to the way we face the world around us (including financial institutions that invest our money and government policies that dictate how our fellow citizens will support us in the future). This sense of an ordered worldview may have a deeper meaning than security in financial terms. The order that underlies security is both personal (as in family relations and discipline) and global (as in the functioning of the universe). So, in this sense, we seek to establish some parallel order in, if not control over, our lives.

Our quest for order as a coping mechanism to deal with the technoscientific culture in which we find ourselves is no different from previous quests for explanations of what is happening around us and why it is happening to us. Whether we choose to listen to soothsayers, priests, or scientists remains our decision, but we expect from them all an explanation of the past, a reassurance in the present, and a prediction about the future. In developing our coping skills, we might sacrifice quite a bit—perhaps too much. So, the question that drives this book is: What responsible balance can each of us keep between accepting too much and too little order in our lives? Put differently, what reasonable level of order might suffice to ensure survival and happiness within the boundaries of an ethical model? Less normative (what you should or shouldn't do) than cognitive (how to think about things working in this world) in its basic approach, this book might help us think through our impressions of the world and our specific roles in it. Yet, this book is normative in insisting that you can and should always think about your choices in ethical terms.

To live in the twenty-first century requires unforeseen skills. Our environment is compounded daily by new inventions that affect us personally. We encounter demands on our personal lives that require mental agility and openness in the face of change over which we have no control. We have to think about what choices we make in relation to a transforming world. The better we learn to cope with these changes, the better equipped we become to retain our composure and sense of identity. In this sense, our mental and physical well-being are intimately connected but at times at odds with each other, setting in place a tension between what we experience and how we cope with these experiences. Technological accomplishments trickle down to daily activities and create situations that we have to critically evaluate (from the use of global positioning systems to cell phones that keep us connected at all times and everywhere, to our places of employment). A critical evaluation is the first step we must take in order to contain and overcome our anxieties and to enjoy the fruits of technology.

We are not alone standing in awe of our rapidly changing environment, our contemporary culture. We can learn from other people's experiences, their ability to take charge not so much of the situation as of themselves. When they are in control of their own feelings and outlook, it seems no situation fazes them in the least. They stand tall and strong, self-assured; they proceed calmly and without much irritation. How can they do this? What makes them seem so strong in the face of difficult and unforeseen circumstances? Admiring their attitude taught me how to change my own, how to figure out what to do and how to do it. I have always aspired to independence, but I didn't realize that the only way to attain it is with the introduction of order in my life as a coping mechanism in a changing technoscientific environment.

It may seem, on the surface, as if independence is the opposite of order, just as freedom is the opposite of discipline. But there is something peculiar that happens when you bring discipline or order together with freedom or independence. It is through discipline and order that the abstract and sometimes vague concepts of freedom and independence become more comprehensible, even easy to recognize. So, instead of looking to sophisticated dictionaries or to scholars in the history of ideas for the definitions, I suggest we look at our own experiences to determine definitions and boundaries that make sense and help us to cope with the conditions of contemporary culture. I found, in my own way, that only with the introduction of some basic order in my life could I appreciate the freedoms in my life. Only when I knew what I couldn't do did I recognize those things I could do, and vice versa.

Now, you may think that this is all too simple, almost common sense, and that everyone knows this. My first answer is, yes—my ideas are based on common sense, but this fact should make them more, not less, compelling. There is nothing wrong with common sense; it is the basis of all the sophisticated reasoning we do as human beings. Moreover, our ability to reason, to be rational in order to make decisions in life, springs from the fact that at rock bottom all our reasoning exercises are based on common sense. This also helps make our reasoning more accessible to other people. Common sense is a wonderful gift granted us by our ancestors so that we can create a human environment for ourselves and for our community.

Everyone knows these basic principles of how we come to make reasonable choices in our lives. If we all use similar tools in making our decisions, and if these decisions can be evaluated by everyone else, then there is hope for constructing a harmonious community of people. This doesn't mean that all people will be necessarily like-minded or that all choices will be the same. If this were to happen, we would be reduced to robots or sophisticated computers. Rather, what this means in the long run is that we have a chance to understand each other without appeal to divine revelation or to a supreme philosopher-king (as Socrates suggested some twenty-five hundred years ago). This is good news! If we can understand each other and the processes by which we come up with different solutions to similar problems, then we can communicate our disagreements and, if we so choose, can even resolve conflicts among people. That's not a bad starting point for achieving not only world peace but also personal peace of mind (as the seventeenth-century Dutch philosopher Spinoza hoped we could four hundred years ago). In short, this is a way to formulate an ethics in the age of technoscience.

So, the title of this book, *Confronting Disaster: An Existential Approach to Technoscience*, is curious only insofar as it could also be titled "The Quest for Order and Freedom" or "The Quest for Personal Peace of Mind." I chose the first because I wanted to emphasize how the quest for order will necessarily lead to the quest for—and even to the fulfillment of the quest for—personal peace of mind in a responsible manner. I must hasten to assure you that I will not be writing an autobiography that traces my own achievement of peace of mind. This would be too presumptuous, as well as inapplicable for many people. My own journey cannot exemplify every person's journey. The paths are too numerous to be reduced to a so-called right path. Moreover, anecdotal evidence can be helpful but cannot guide everyone. Our experiences vary too much, and the circumstances we encounter in life are too different to be distilled into one picture or model. So, I retreat back to the history of ideas, the stories and theories told to us by our Western ancestors. They look for common threads and principles that may be applicable across cultures and years, that would provide the kind of coping methods by which some harmony and ethical demeanor can be guaranteed.

As a college professor, I have the advantage of having read some of these historical texts and of having tried to explain them to many students over the past fifteen years. But this seeming advantage also has a major drawback: I have been confined to the halls of the academic, insulated world. The language spoken there is remote, esoteric, sometimes totally incomprehensible!

Therefore, I also draw on my experiences in the business world, where people speak more directly to each other, express their emotions more openly, and have no problem dismissing any authority that is in their way. The mark of a democratic society is the critical expression of every citizen. This principle translates itself to the workplace both for employees and for consumers. We demand to know why a product or service is priced the way it is. We expect to understand how the system works, even when it works against our will or desire. We think we have the right to know, to question, and to receive appropriate answers to anything that we come across in our daily lives. In short, we expect to find reasonable order in our human-made world. Incidentally, this is also true of our expectations of the natural world, and it is why we are so obsessed with scientific knowledge. Tell me why the sun rises every morning. Tell me why the weather changes the way it does! While we forgive economists for their inability to explain, let alone predict, the fluctuation of the stock market, we expect scientists to answer every biochemical question with the certainty (but unfortunately without the humility) of an Einstein telling us about relativity theory in physics (at the early part of the twentieth century).

It's interesting to recall that when gamblers tried to figure out how to improve their chances in casinos and merchants in the sixteenth century tried to control the odds that the goods they shipped overseas would arrive safely, they turned to mathematicians and scientists to construct models and principles (probability). Our expectation to find order in the world transcends nature as such and filters into human relations and the way we organize our political world. Because of this ongoing expectation I wander from science and technology to the affairs of the state and to the personal psyche of humans. I believe that at some level there is an expectation about peace of mind (or happiness) that is connected to all of these areas and that can be fulfilled through coping strategies that are ethical.

For all of these reasons, then, I try to bring together broad cultural experiences of the twentieth century and relate them to the way people think about their daily lives.

My presumption here is that Sigmund Freud's *Civilization and Its Discontents* (1930) is a wonderful model for what I'm trying to say here, since it moves from the individual to the community at large without sacrificing the personal concerns of people for the sake of a larger social setting. Likewise, Viktor Frankl's *Man's Search for Meaning* (1946) is a wonderful model that moves from the horrors facing the Jews during the Holocaust to every individual's quest for meaning. Though Freud provides impressive parallel pictures of civilization and the individual psyche as an explanatory model, Frankl's prescription is more informative as a guide for future action. I agree with both of them and many others who write in this fashion that large, cultural phenomena are intimately linked with and influence the way we feel, think, and behave. In other words, individuals tend to internalize the cultural cues they observe, whether it means policing themselves to conform to social norms and moral codes or limiting their aspirations. There are certain things we take for granted about our expectations (health care, jobs, careers, living conditions, personal safety, and national security), without realizing the social, political, and economic costs we must pay for them to be fulfilled. In this sense, there is no separation between the private and public spheres, as so many political theorists insisted earlier in this century (e.g., the philosopher and political theorist Hannah Arendt).

In some ways, this book provides a brief explanation of how we have come to our current situation at the beginning of a new century. In other ways, this book deals with the kind of mental, psychological, social, and personal coping strategies we can develop to cope with our anxieties about the conditions that determine our success and failure. In yet other ways, this book can serve as a manual for how to think about a more fulfilling life, not in an abstract way but in terms of the limited order we can establish responsibly on a daily basis. For this purpose, this book is a cultural critique of contemporary Western society, with its technoscientific zest and promises of freedom. I consider it a necessity and not a luxury to think critically about the conditions of our existence, because deep down I believe that we can control our emotions with our minds (or at least reduce the pressure of incomprehension). Critical consideration is useful for finding explanations about the past as well as for providing alternative courses of action in the future. Once we understand what options we have and recognize the criteria underlying our choice among them, we will be better equipped to recognize success and admit failure. Regardless of which way it turns out, our process of deliberation can be construed as ethical, because we take into consideration variables outside of personal greed, fear, and self-interest.

There are, of course, other examples from the history of ideas that inform my own approach, and I would like to contextualize my effort within this framework. Maimonides wrote *The Guide to the Perplexed* in the twelfth century, and E. F. Schumacher wrote in 1977 a book with the same title. While the former is concerned with biblical interpretations and the life one ought to live in accordance with divine law, the latter is concerned with the centrality of our existence in a fast-changing world in which environmental issues are just as important as the relationships among people and nations. In between the medieval religious concerns and the concerns of the late and post-1960s, there are a few texts that insist on examining the human condition in the twentieth century in the terms dictated by the horrors of World War II. I already mentioned Frankl's book, based on his own experiences in a concentration camp and on the psychological implications of that experience in terms both of its effects on the individual and of its therapeutic implications for an entire society. The other is Simone de Beauvoir's *Ethics of Ambiguity* (1948), in which she takes issue with other existentialists of her time. For her, the engagement with one's own sense of ethical ambiguity is a rallying cry for passion, empathy, and action. In all of these texts, we can detect the residues of Epictetus's ideas (circa 50 to 130) concerning the principles of stoicism more than in John Dewey's *Quest of Certainty* (1929), in which American pragmatism is invoked.

What can be expected from this book? While I try to speak plainly and to express some interesting ideas in simple terms, I'm sure I have the handicap of someone who has written too long for academics and specialists. Having said this, I also believe that there is some value in pondering the ideas of scholars of the past, and pondering them on their own terms—with all the trappings of a subtle and sophisticated language. If the language of intellectuals helps sort out our subtleties, then there is nothing wrong with it. If, on the other hand, it is used to intimidate and ensure that only a few can master it and so expose its flaws, then it is wrong and shouldn't be tolerated. My attitude toward intellectuals is no different from my attitude toward musicians and athletes: Even the most difficult music or athletic activity can be explained, even taught, to anyone who is willing to take the time to learn and train. This doesn't mean that anyone can accomplish as much as the experts and stars; rather, a person who attempts can learn enough to appreciate the accomplishments of the virtuoso or Olympian and the processes needed to bring them about. Concerning the language of intellectuals, common sense, as I said earlier, still rules! Fancy words that mask silly ideas should be exposed. Fancy words that bring together interesting ideas should be adored! Back to my sales pitch for this book.

First, this book will try to illustrate why we feel confused and anxious about our position in the world by connecting us with the forces of history. This means, in plain English, that I will spend some time reviewing certain developments of the twentieth century in science and technology so that we can better appreciate their influence on our daily lives and personal attitudes in regard to the disorder that we experience. Second, this book will also explain why we feel the need to find some level of order in our lives. This is not a disease or an obsession but a normal response to the conditions of our lives. Third, this book will try to shed light on how to lead a more balanced life so that anxieties do not turn into paralyzing anguish and depression. In short, this book is a call for action, an empowerment tool with which to conquer our anxiety and to express our involvement with our social and material surroundings. So as not to mislead anyone, though I use the language of feeling here and there, I don't pretend to be a psychotherapist in the traditional sense, Freudian or otherwise. Instead, I insist that how and what we think influences how and what we feel. My approach, therefore, is cognitive in nature and is based on my conviction (which itself dates back to Spinoza) that our thinking can and does directly influence our feelings, that we can control our emotions and direct our energies thoughtfully and critically.

CHAPTER TWO

The Transformation of the Human Condition

Technoscience and the Conditions of Humanity

"Technoscience" is a term used in the late twentieth century to describe the intimate connection between science and technology. No longer is science understood to be a theoretical endeavor in contradistinction to technology as applied science. Neither is it understood that science has a chronological priority over technology—that is, first scientists think of great ideas and construct theoretical models that eventually see their experimental application in technological innovations. Rather, there is an appreciation that technical tools are instrumental in bringing about scientific discoveries and that they are essential for the scientific enterprise. This appreciation can be traced to ancient times, when navigational instruments enlightened sailors and scientists to appreciate celestial movements and the relationship between the phases of the moon and the tides. Likewise, the telescope helped usher in a new set of models concerning the rotation of the earth and the circular (or elliptical) positioning of the stars. Lens grinding and the microscope were essential for discoveries regarding human physiology and the immune system.

Technoscience, then, is a shorthand designation of a changed worldview concerning the evolutionary (and at times revolutionary) process by which our knowledge has expanded and changed over the years. It is a term we use more commonly today to acknowledge the debt theoreticians owe their fellow engineers in the production of knowledge claims. It is also a way to describe not only what we know but how we learn what we know. This process was popularized in the twentieth century in the cases of "big science," such as the Manhattan Project. While congregating in Los Alamos, New Mexico, scientists and engineers worked together to develop the atomic bomb, losing a sense of their separate identities as scientists or engineers. They thought through the nuclear physics of their day while getting firsthand exposure to the technical apparatus with which they worked, experimenting with the materials that would enhance or retard the fission of subatomic particles.

On a more personal level, the effects of engineering wonders seem, on the face of it, positive and revolutionary. Changes in transportation and communication have altered our attitudes and commitments toward both our personal and professional lives. We fly to visit relatives and attend family functions as if all in the family lived next door; we call overseas at a touch of a button—a pleasant closeness that would have been inconceivable only one hundred years ago. We can drive quickly to comfort the sick and save the lives of dear ones and strangers alike. We can alert millions at a stroke of a few keyboard keys that a danger is imminent—a flood, hurricane, or drought.

Yet, technoscience exacts a price: It promises freedom and access to the world at large, but the World Wide Web includes new forms of insufferable commercialism, bondage (addictive appeals), and surveillance. As soon as you click on your desired icon, a different icon or advertisement pops up and offers you yet another item or service you really don't need. We have become codependent on the computer for entertainment over the Internet, for passing time playing card games, and for taking care of our daily chores. Also, as we explore the world and connect with its numerous facets, we have found out that somewhere someone (or something) is keeping track of what we do. This could be a government agency or even Amazon.com compiling our profile, once we've ordered books or music through them, or our employers, who know when we log in and out and where our Internet adventures have taken us in the past day.

Technoscience as applied to health care promises longevity along with sophisticated medication and clinical treatment of almost any disease known to us, but it also delivers medical intrusion and financial hardship. For example, we can preserve (this term might be too strong) human organs and sustain them with machinery for a long time. There are ways to keep our hearts pumping blood and our lungs circulating oxygen to the body for weeks, months, even years. There are ways in which we can "clean up" kidneys so that toxicity from alcohol consumption is mitigated. Doctors have been able to save smaller and younger fetuses in incubators over the past two decades. But at what expense? Who pays for the improvement in medical technology and the advancement of medical research? What other costs are deferred for this purpose? What can we not afford to repair in the meantime (roads, bridges, broken households of the very poor)?

As the prevalence of technoscience in our daily lives becomes more pronounced than ever before, especially in terms of the negative side effects or consequences of its implementation, there is reason to approach more critically any positive claim about the transformation of the human condition. Haven't we in our culture made a type of a Faustian deal with the gods of technoscience? We offer complete obedience and adoration of everything scientific and technologically innovative: computers, cellular phones, high-powered surgical procedures, and all the other gadgets we have learned to enjoy in return for full acceptance of the technoscientific way of life. Even when something commercially viable poses a threat, such as nuclear energy or automobiles, we appeal to science, technology, and engineering to find the solution to the last technoscientific danger or disaster. Our deal with the gods of technoscience is never to fully or overly challenge their supremacy and their permanence in our lives. We won't just give up on our instruments of joy but only ask that they be converted, transformed, improved, and made to serve us more gently. In this sense, then, we appeal back to them to resolve whatever problems they brought about in the first place. We return to them as the final arbiters of what needs to be fixed and taken care of, and we are willing to expend whatever funds are needed to support their efforts. When they are successful, we are grateful; when they fail, we understand that it's our fault for endorsing their products to begin with. Either way, we end up paying for the results.

Put differently, when we give in to the gods of technoscience and their generous gifts, when we open ourselves to accept their gifts unquestionably, we put ourselves in the precarious position of accepting much more than just these gifts. We accept a way of life, a way of conducting ourselves toward others and the environment. We open ourselves up to abuses that might never have been intended but that have serious effects that are unavoidable: surveillance, lack of privacy, removal of boundaries between the public and private domains, complete exposure of our secrets, and other effects the potency of which isn't visible for years (for example, genetic alteration of foodstuff and its effects on the immune system).

From health care to the Internet, technoscience is inescapable. Science and technology have influenced, even transformed, not only the conditions under which we live but also the human condition. What this means exactly and how we must adjust to this phenomenon are explored in more detail in this chapter. The transformation of the conditions of humanity has been quite radical in the past century. We moved from not considering the consequences of polluting the environment and cutting trees, for example, to an environmental sensitivity, a holistic appreciation of our human position and role within nature. Our behavior displayed what I shall call a detached engagement with nature. By "detached engagement" I mean the way we have looked at natural resources as something we can use as much as we want without thereby being changed ourselves. This means mining natural resources like coal, oil, natural gas, and even chopping trees. What difference, we might ask, does it make if we take what is around us? How will this change our lives? Obviously every time we take anything away from our environment and transport it anywhere, we pollute the air we breathe and exhaust a precious resource. The detachment we assume when we do this is that the environment is out there, while we are right here. Being "out there" meant that it was available for the asking. All we needed to do was take as much as we wanted, and nobody would or could stop us. We even had an entire ideology and folklore around this kind of activity: exploring the wilderness, wandering into the West, manifest destiny.

The fact that we and the environment in which we live are intimately connected and the fact that the one influences the existence of the other both seemed to escape us, unless we were faced with some catastrophe (spillage, pollution, accidents, nonrenewable resources, even loss of human life). Perhaps what lurked in the back of our minds was the biblical story of creation and the sense that we, as humans, were destined to rule the earth on which we lived. Perhaps this seductive narrative appealed to our sense of superiority; we believed it wholeheartedly. Perhaps we never quite understood what Native Americans meant when they told us that they were custodians of the earth and that they have to ensure its continuous prosperity (despite our recent arrival on their shores).

It's true that over time we have realized a holistic appreciation of our human position within nature, so that we have become more environmentally sensitive to the damage we cause when we pollute or cut trees, for example.

I bring up environmental issues at this juncture to give an example and draw a parallel between our changing approach and technoscience. A holistic attitude is warranted in both cases. Just as technoscience has engulfed everything we do, so has our interaction with the environment. We no longer can view either as something out there, remote from our daily lives, but rather as something in which we participate, that we affect. We are no longer distant observers who merely study what happens to us but rather active participants who affect changes that are in turn affecting us. We cannot afford to claim that somehow technoscience is independent from our actions or that nature continues to evolve independently from us. As we appreciate our roles and the power we exert over time, we must become more responsive and responsible, sensitive to the levels of control we have over our dominions. This means, for example, realizing long-term consequences that we might deem negligible in the short term.

The tools of technoscience are within our control and impact our behavior and the way we live. Instead of believing that technoscience is only a by-product of our actions and a tool for our survival, we are more aware that technoscience has an immense impact on what we do and how we go about doing it. For example, transporting natural resources for energy use across the continent (not to mention importing them from across the globe) lets us forget how precious these resources are and what effect their extraction has on the immediate environment from which they come. In many ways, we have learned to appreciate that our reliance on technoscience could have adverse effects on nature. If nature is part of the context within which we operate, then technoscientific tools should be used in a manner that protects natural resources and ensures a necessary equilibrium we ought to maintain. In short, nature and technoscience must coexist harmoniously rather than conflict.

We seem to forget that the tools we use have become the focus of our attention and the main source of our self-expression, if not our identity. People speak in the United States about their houses and cars, about their material toys, as if they define their very being, as if they sum up who they are and what their goals in life are. But these remain instruments, extensions of our hands and brain power; they cannot describe who we are as humans or how we are connected to the environment in which we live. We remain part of the environment and depend on its nourishment for survival. So, in order to interact responsibly with nature, in order to enjoy its fruits and gifts without spoiling it, we need to use our tools efficiently and carefully, thoughtfully and sparingly, so that our enjoyment doesn't turn into a hellish nightmare.

Though we are accustomed to thinking that by the time anything major could have an effect on our lives we would be long gone, we have seen in the twentieth century that, the time span of these changes become shorter and shorter. Some of what I'm speaking about isn't originally presented as being detrimental at all. Building enormous dams for the generation of energy in remote desert areas such as Las Vegas, Nevada, is presented as an engineering achievement without major drawbacks. But eventually water use and wildlife are affected, and some twenty or fifty years later we come to find out that there are unexpected results that we wish were otherwise: water shortages in the highlands, extinction of fish along the river, and the decrease in irrigated farmland along the way.

The realization that the technoscientific mind-set may be in conflict with nature, that it may provide opportunities for conquest and empowerment, and that it may set us free from the natural limitations we resent (about survival, life expectancy, and mortality) dates back to the eighteenth century. The realities of the Industrial Revolution, for example, have transformed how we think about our role on this earth and how we use natural resources. Though we might resist the comparison, the twenty-first century is as much indebted to the eighteenth century as it is to the twentieth. The evolutionary development of humanity and its quest for the material improvement of its lot is at least three hundred years in the making.

The Enlightenment ideals of the eighteenth century set the tone for progress and the improvement of the human condition. With the shift from superstition to experimental science, from divine revelation to critical inquiry, every human being was encouraged to pursue a better life and was promised that happiness could be attained. Of course, some would argue that this was a false promise, since with the freedom to pursue comfort and happiness came a level of freedom and responsibility that eventually increased anxiety. But the promise was still there, offered as a challenge for the conquest of nature and the empowerment of humanity. Instead of appealing to an invisible God, European thinkers and political leaders learned to appeal to the gods of technoscience, theories and practices that would transform our material conditions. Unlike the appeal to God, it was felt that the leaders of technoscience, its high priests, would be accountable and respond directly to our pleas. What's missing from many present accounts of the human condition is the appreciation for this radical transformation in the human psyche. The sense of liberty and personal empowerment involved in this transformation is momentous, almost unbelievable. To some extent, in many European countries humans were not perceived to be, or given the credit for being, the center of attention, the center of the universe, or the center of power relations (up until the age of Enlightenment).

For most of history, people were set in communities and organized socially and politically in a fashion that left them at the mercy of powerful institutions and the benevolence of kings and priests. The hierarchies of the church and state were structured in ways that limited mobility and self-expression, so that most of the population was resigned to occupy the same spot their ancestors occupied. This meant that you had to be content with your lot, accept what was handed down to you, and be relieved that life would move along at a predictable pace, with predictable results. If you happened to disavow your position or demanded to change your situation, you could be ostracized or even excommunicated.

For some it was the awe of the divine that kept them humble and obedient to the church authority; for others it was the fear of the wrath of a powerful king and his vicious agents of terror. Even Plato's *Republic* (of ancient Greece) was a political model that prefigured where everyone belonged and what everyone was supposed to do in life, despite its benevolence and high-minded concern with the general good of the community (in some democratic sense). Under these circumstances, we shouldn't be surprised that for so many generations most of the population remained stable (some would say stagnant), almost frozen in its position. Custom and habit, tradition and ritual, all conspired to retain a status quo that benefited those in power, ensured low self-esteem among common men and women, and undermined any sense of ambition for the improvement of one's lot in life.

Obviously stability is comforting and can be construed in a positive manner, for it allows one to know what is expected of her or him and what rewards might come her or his way. But that level of comfort limits one's aspirations and the freedom with which to pursue novel ideas or unexpected ventures. Stability nurtures the maintenance of power relations and sets limits on one's dreams of change. As our horizons close on us and restrict our activities, they also dash our hopes to create our own destinies. When we believe that the boundaries of our activities are preset, we become less motivated and less responsible, for we don't believe ourselves to be instruments of transformation. Herein lies a great loss for future development and personal satisfaction.

Progress and Pressure

The Enlightenment pushed aside some of the traditional notions of order and opened the field to more players with more options for personal decision making. That meant that predestiny, for example, or the acceptance of one's station in life (the ordered universe according to the church or the king), was open to challenges or personal reconfiguration. With this, of course, came the pressure of deciding what to do in life, the anxiety of choosing among alternative options, and having to bear the consequences of the decisions that led to them. In short, with personal empowerment comes personal responsibility; with responsibility comes anxiety; and with anxiety comes a quest for stability and order. If you make your own choices regarding what to experiment with sowing on your land, then the results of this choice—let's say, disastrous crops and financial ruin—become your problem, your fault. You can't turn around and claim that you simply obeyed your king, that it's his fault the land yielded poor crops. This process of having to deal with personal choices amid technological innovations and the prevalence of the scientific dominance of everyday life is fairly new. Though we have freed ourselves from so many shackles along the way, deep down we return to the quest for order and stability, as if we never left the infancy of our Western history. We want someone else to make the hard choices; we want someone else to be responsible if anything goes wrong. Of course, when success is at hand, we want to get credit. But it's not quite clear how often we succeed or how we measure our own personal success. The log book we carry in our minds in regard to our success and failure rates needs to be organized, so it shows us how our world is ordered and who is behind the order imposed on us. This is how, even in the age of technoscience, we view our personal situations in relation to the community at large, even the entire world.

Perhaps the parallel that Freud draws between the psychology of the individual and that of society as a whole is useful here. Just as the infant seeks the warmth and protection of a parent, so does a civilization. The progress of civilization can be perceived in terms of the progress an individual makes over time, moving from infancy to childhood, from adolescence to maturity, from adulthood to middle age, and from there to old age. The analogy is difficult to sustain, because we have to illustrate to what extent a civilization matures (like a human being) and to what extent it reverts to its childhood anxieties in old age (like a retired person who worries about physical and financial security). But however frail the parallel, it does help elucidate the key or critical issues in the transformation of the human condition.

As the twentieth-century social critic and cultural theorist Christopher Lasch reminds us in a different context, the idea of progress makes sense to us and can be sold to future generations only with a switch from a culture of scarcity to a culture of abundance (1991, 78). In a culture of abundance, we believe we have more stability and security, as the Garden of Eden was portrayed. I bring up the biblical narrative of the Garden of Eden as an important ingredient in the formation of the Western psyche, as a narrative that influenced the way we have learned to think about the shift from abundance to scarcity (the Garden-to-Fall scenario). Adam and Eve had no worries regarding their needs for survival! The Garden was plentiful, and they could roam and enjoy it without concern for future plans or potential disasters. Once we move to a culture of scarcity, in which we believe that we won't have enough to survive and that in order to survive we have to compete (and thus feel less secure), we live a life filled with worries.

But there is something perverse in the switch from the psyche of abundance to scarcity that is only marginally connected to the Bible and the story of the Fall. It's unclear if this switch has been used to explain our more contemporary human condition at the dawn of civilization or as a way to inspire us to dream of a return to a Garden of Eden–like existence. Has the Garden-to-Fall narrative been a warning we must heed in every age, or has that been understood as a justification of current circumstances? It seems as if the notion of scarcity is the foundation of our current belief system about our existence (even environmentalists endorse it), and to a great extent it justifies the capitalist model of market forces. Perhaps environmentalists, for example, worry that we would conserve less or abuse natural resources more if we believed them to be renewable by definition and therefore abundantly available. Perhaps even they use the notion of scarcity to affect our thinking and our worldview.

If there isn't enough to go around, then we must compete for resources that will help us survive. These resources—land, crops, water, minerals, underground ores, animals—are all limited in their availability, and therefore we have to learn how to outmaneuver our neighbors and friends, in order to have more than they do. This is a "natural" justification for the competitive attitude we have toward everyone else in our community. I can't help it, we say, because if I don't have more and hoard more, then my neighbor will get it all, and I will suffer (or in the extreme case, starve to death). In the age of technoscience, as opposed to the Garden of Eden or even the Enlightenment, things should have progressed beyond this view, but they haven't. Our society has believed (on some level) that technological innovations would ameliorate our situation and avoid the fear of scarcity. But this hasn't happened. I would maintain that the reason for this has more to do with a mind-set—a psychological disposition—than with the actual situation in our environment.

If there were abundance, then the supply of goods would be, in theory, infinite, and regardless of the demand for these goods, prices would be relatively low. If this were the case, everyone could afford anything, and the feeling of abundance would be commonly shared (and the fear of scarcity would vanish). To put this in economic terms, prices rise due either to a rise in demand for goods and services while the supply of these goods and services stays constant or to a steady demand while the supply is decreasing. To justify high prices, to justify the profit margins that accompany high prices, there must be a solid assumption about the scarcity of goods (natural resources and all the manufactured items that rely on them). The dual concepts of abundance and scarcity, as well as the notion of progress, stem from the tenets of classical capitalism. The notion of progress, in turn, is related to a modern drive to compete for survival and success. But along this road to success there are many obstacles, which produce a great number of worries. These worries, in turn, reflect a sense of anxiety and insecurity. How can these human emotions be dealt with? What countereffects can be produced in the human psyche?

Whether it is found in the philosophers John Dewey's work *The Quest for Certainty* (1935) or William James's *Pragmatism* (1908), there is a "childish desire for certainty," as Lasch calls it (1991, 289). This desire is for certainty regarding the material conditions that provide the personal prosperity people experience in their daily lives. Lasch is probably correct in this assessment (both about the childish nature of the culture and about its desire for certainty), because all the indicators we had by the end of the twentieth century were of a culture as concerned with its psyche as with its material well-being. The psyche is not something so "sick" that it desperately needs the cures of psychotherapists; rather it is in search of comfort and security, meaning and happiness, in a changing world that is becoming more and more complex for our limited understanding to grasp. But how does the idea of progress fit into this scheme?

On the one hand, there is great promise and hope, and the possibility of ensuring a life infused with meaning for all of humanity. To put it bluntly, there is the prospect of transforming the human condition (replacing anguish with pleasure and hope), improving our position in life, and empowering humans in a world that is friendlier toward their needs and wants, that cares about how humans feel about themselves and others. If there is enough to go around (a sense of abundance), then there is no need to compete and fight, to worry about the survival of the fittest. In this scenario, technoscience can be instrumental in getting us there, and the more we have of it the better. At least this was the view of those promoting the scientific revolutions of four hundred years ago, and those who brought about the Industrial Revolution of the nineteenth century. For them, there was a great promise that could be achieved through the careful implementation of technoscientific models, whether they related to agricultural practices (from the cultivation of land to cattle breeding) or transportation (from bridges to the steam engine).

The utopian vision of progress is what led Karl Marx to envision a communist ideal in which everyone would work enough to satisfy herself or himself and the rest of community, while enjoying the fulfillment of personal growth through education and the arts. Utopian thinkers throughout history assumed a level of abundance that would take away the worst of our human traits (fear and greed) so as to cultivate the best (empathy and generosity). The belief was that if you had more than you needed, fear, greed, and competition could be relaxed and make room for kindness and altruism.

On the other hand, the idea of progress failed to account for a different kind of pressure that would be imposed with its manifestation. It's not the pressure of material survival (meeting one's basic needs) but a psychological pressure (trying to keep up with progress itself). We feel the need to get better every year, to move forward whether or not we like to move at all. We have to keep up with new ideas and innovations so as not to fall behind, or even be crushed by those who embrace progress. The advances of technoscience must be seen in this light to be positive and worthwhile, supportive and essential for bringing us closer to a utopian ideal regarding abundance.

What has radically changed over the past century (and something envisioned by the Enlightenment leaders and their critics) is that the pace of change has become more rapid than ever before. Transportation and communication are two areas in which rapid changes are most noticeable—the horse and buggy were replaced by the automobile, and then the airplane, and then, who knows, even spaceships; the mail moves more quickly with the Internet than with regular carriers (or pigeons). The cell phone is now connected to the Internet and the facsimile machine so that a simple conversation is now enhanced, recorded, and monitored in a way inconceivable only fifty years ago. One consequence of this progress is that corporate leaders report their revenues and earning quarterly (as opposed to annually) and face the judgment of financial analysts and stockbrokers (not to mention their own shareholders) immediately and constantly. Students are graded quarterly or every semester and are constantly assessed with a battery of tests, as opposed to waiting till they graduate from their courses of study and are tested in their chosen fields of research and work.

Our culture has learned to measure and assess its surroundings in extremely small time increments, so that a year seems long and its end far away. This is tantamount to a shift in worldview, a conceptual transformation the likes of which happens only every few hundred years. For example, the old-fashioned Soviet five-year plans, in which the Politburo set goals for the entire nation that would be fulfilled within five years, gave a sense of planning for the future (and the time frame was long enough for a long-term view). Elective political offices in the United States, such as the presidency or the Senate, are held in four- or six-year increments, respectively, so as to allow enough time to elapse between one's plans for the future and the time needed to accomplish them. But to look at ourselves in terms of a decade or a century seems unreasonably long in duration, as if we would be lacking crucial information along the way. The wait-and-see attitude of previous generations is less appealing in the modern age, when time increments have been shortened to television sitcom shows, or what politicians call "sound bites."

The respect for one's knowledge that is earned over a lifetime of accumulated experiences and information has been replaced with an expectation that one's knowledge is earned with a college degree in four years. Wisdom is a life-long learning process that requires knowledge and experience, engagement with nature, other humans, and books. It stands to reason, then, that if knowledge must be accumulated and critically evaluated over time, one's age becomes a factor in appreciating one's wisdom. Therefore the Jewish law forbids anyone under forty to study certain texts. No matter how brilliant and mature someone appears, years of living among others, and years of encounters with success and failure, are required in order to engage certain ideas that would otherwise seem overwhelming.

Perhaps the new worldview of smaller and smaller time increments is reasonable in the face of disasters that could have been averted if action had been taken sooner. For example, instead of waiting for years to observe (and suffer from) the side effects of strip mining in the Rocky Mountains, we should right away have examined the soil and checked how certain chemicals affected the water and the environment. Perhaps it is reasonable to make us and others accountable for everything we or they do. But does this mean that we have to lose sight of the big picture? Does that mean that we penalize an individual or a corporation for losing some ground in the short term as they invest time, energy, and money for the future, rather than giving them the leeway to show the payoffs in the future? Only a long-term perspective can account for and appreciate immediate sacrifices for eventual benefits. This is true of education in general (when years of studies don't seem to yield immediate results), and of technoscience in particular (when "big science" projects take years to complete, with enormous funding demands and no observable benefits).

Perhaps another way of explaining the predicament of evaluating the conditions of humanity and the stress that accompanies progress is to stress the complexity of measurement. Without measurement, assessment is impossible, for there are no standards, baselines, or criteria according to which to evaluate rationally. One can always evaluate irrationally—that is, be emotionally ready to judge something in an arbitrary manner. But if we are to follow the Enlightenment injunction to be rational and be open to critical examination of our own evaluations, then we ought to lay out our criteria of assessment. Constantly measuring everything we do and providing smaller and smaller time frames for such measurement limit our perspective. When our focus narrows, we fail to see the big picture and how the data we measure are related to others that might alter them (and therefore alter our own measurement of the original set of limited data). Our measurement, then, become less informative and less useful and might increase, rather than decrease, the certainty we wanted to attribute to them. In other words, instead of helping us gain knowledge and peace of mind from what we analyze, the criteria make us more confused and less certain about aspects of life we wish to understand.

Because short-term assessments are foremost on our minds these days, measuring audience reactions to television shows and newscasts within minutes of seeing them, the progress of the information and technoscientific age comes with pressure. Pressure means added anxiety that worsens our human condition. We are more fearful, more vulnerable and suspicious, more concerned about how to justify our immediate situations in the face of an uncertain future. Here is the rub: While our material conditions may have improved and our relative wealth frees us from toiling from dawn to dusk in the fields of our forebearers, our inner lives are in more turmoil than ever before. We have anxieties our grandparents couldn't have fathomed. For example, Americans worry about political, economic, and military instability around the globe, because any incident may affect energy prices and the economic conditions of our country. If the president decides to send troops overseas, the military budget may need to increase, and our personal taxes may increase as well. This cycle of events is both national and international, as the term "globalization" has come to be used. Part of the problem, so to speak, is that we know more about what happens around us and that we are informed more often about incidents that happen around the globe. The knowledge itself has exposed us to the anxiety and emotional vulnerability I have spoken about—the more we know, the more we worry. The more we worry, the more personal pressure we feel to change the world, or at least our immediate environment.

We worry about currency fluctuations in financial exchanges with names we cannot pronounce. We are concerned about them because they may affect our own interest rates and therefore the future costs of our homes, cars, and all other items we buy at the store. To some extent, we are back with the eighteenth-century philosopher Jean-Jacques Rousseau and his "noble savage," whose nobility overshadowed the jungle where he lived. His survival was related to his ability to negotiate between the threats and good will of others. When the noble savage displayed pity, then "the survival of the fittest" didn't matter. Pity, as Rousseau so eloquently reminded us, was the first indicator of humanity, a way to relate to other humans beyond the aggression and cruelty we associate with the notion of the survival of the fittest. I bring up this notion because fear and anxiety became excuses for acting out in the late twentieth century, when competition became a way of life and greed a way to avoid human civility. By "acting out" I mean the sort of behavior people display in public spaces (restaurants, taxi cabs, trains, and airplanes) that they justify after the fact as necessary in order to get anywhere in life. This could be pushing other people to get ahead of the line, ignoring other people's pleas for help (with a heavy suitcase or with patient conversation), or being unnecessarily aggressive in human interactions.

In the age of progress, when the rapid pace of technoscientific changes stares at us through television sets and the print media, we must take action. Our actions are not necessarily outward looking but have much to do with how we conduct ourselves and take control of our emotions. We have to realize that many of the events we hear about we cannot change or control. But this doesn't mean that we avoid knowing about them or dismiss them as irrelevant. These events are relevant to how we feel about ourselves and how we cope with them personally. If the world appears cruel and competitive, hateful and unforgiving, we must stop and examine how we have come to this point, what have we accepted along the way to conform to this particular worldview.

We must realize that to be responsible is to be responsible for our own behavior toward those closest to us, in light of a situation that was set by others. If we feel pressured, we must look for the antecedent conditions that brought it about, whether in terms of technoscientific developments or in political and economic terms. We can learn about the past so as to appreciate our present and plan our future. As we learn about the conditions of humanity, we have a chance of learning about the human condition without losing sight of our human dignity and the dignity of others. Also, as we progress through life, we must share our own lessons with others and ensure a more hospitable environment in which to raise our children. I think this will go a long way toward the ideals of nobility expressed by Rousseau. This is not the nobility of the aristocracy that claimed status and power but a nobility of soul and heart that can soften the blows of incessant progress.

Relationships among humans must be based on something other than self-interest and the pursuit of individual goals. They must involve empathy and sympathy, compassion and sorrow we feel for each other. We are humans after all, regardless of any model of progress or the promises of modernity. I can see why the romantics of every generation rail against the pursuit of progress and modern technology. It is not because they refuse to enjoy the fruits of progress or the benefits of modern technoscience, from improved health to safe workplace. It is because they worry about the high costs of this enjoyment, the price that we exact from each other at every stage of our self-development. Environmentalists look at this situation in terms of the desecration of some idealized notion of pristine nature; religionists worry about the loss of spirituality; and poets decry the loss of soul. They all point to the same predicament: Are we willing to lose more than we gain by buying into the heightened drive to reach new heights of material enrichment?

Some would say that the predicament I'm describing here is misguided, almost false. Why must we speak of a price when none is exacted from us? Why must we set up a predicament when in fact our situation has improved over time? Of course, the material conditions of humanity have improved tremendously over time, if we only consider our longevity. In the short time span of the past century, the life expectancy of American males has jumped from forty-seven to seventy-four years. This amazing increase includes variables relating to health and issues ranging from hygiene to labor conditions, from medicine to wars, from living quarters to communication. So, why complain? We should celebrate our achievements, and yet we can't shake our internal unrest. What is it that drives us to seek control over our environment and find order in our lives? What holds us back from this celebration? What makes us anxious when we should feel free and calm? What keeps us awake at night, when most of our basic worries have almost disappeared?

Expectations and Aspirations

Perhaps we take too much for granted; perhaps our expectations have changed over time. Or maybe the threshold of satisfaction and security in the Western world has risen and we have lost sight of what should and shouldn't matter. Perhaps what bothers us deep down is our own culpability and the responsibility we feel toward ourselves and others. We cannot blame the world or particular individuals in it for causing our anxieties and hurting us, since we are all participants in determining the material conditions that affect us. True, some of us have more influence and power to bring about change. But this doesn't mean that even our complacency is immune from criticism. Our basic needs are met most of the time (at least in the Western developed world), so we have confused the distinction between needs and wants. Karl Marx was clear about the distinction between needs (necessary conditions for survival) and wants (socially constructed and fetishized desires) and about the evolutionary changes that go along with that kind of classification. He was also aware that we could manipulate this change with enough marketing and advertising savvy, so that an expensive sport utility vehicle could conceivably become a need and not a want. Where do you draw the line? More interestingly, who draws the line? When the line is drawn, whose interests are being met and satisfied?

Once Marx set the tone for this line of questioning, the whole naturalistic approach to the human condition was challenged. No longer could we naturally assume a permanent set of needs and wants, as if they were God given, natural. Instead, our human needs and wants have become culturally contingent, socially defined, and politically established. They are culturally contingent because new products are introduced daily, so what we thought we didn't need yesterday becomes our need today (cell phones). Our Western culture defines itself in terms of the technoscientific progress it brings about; it is an ambitious culture that expresses its superiority (at least in its own mind) in material terms. Our needs are socially defined in that social conventions and peer pressures change what we are supposed to like, dislike, or crave (tanning salons and weight-loss diets). Our Western society is interwoven into a global web (or maze) of material achievements and the means by which to measure ourselves through the acquisition of material comforts. They are politically established because there are laws to limit our desires (pornography) or sanction those that are culturally introduced and socially established (state-funded lotteries or gambling casinos).

To question any worldview or opinion requires a critical ability (trained skill) and a will to do so. The seventeenth-century French philosopher René Descartes understood this as an invitation to establish certainty through the method of radical doubt. If you want to know something for sure, don't just accept it or take it for granted because generations of people claimed it to be so. Rather, doubt it completely and thoroughly, and then, bit by bit, find out how certain you can be of the claims you examine. Radical doubt is the first step toward certainty, toward building a secure base of knowledge on which you can depend and from which you can venture into future studies.

In the twentieth century, the American philosopher John Dewey echoed this sentiment and approach, and put it this way: "Any philosophy that in its quest for certainty ignores the reality of the uncertain in the ongoing process of nature denies the conditions out of which it arises" (1960, 244). Perhaps Dewey was right to be concerned with the quest for certainty as a way to deal with rather than ignore the inherent uncertainty that looms in our daily lives. We have marks against us from the moment we're born. We have no choice as to when we appear on this planet or who our parents are. Yet we have to take full responsibility for our very existence despite these fundamental choices we have never consciously made.

Think about it: The most fundamental choice is not ours! Many religions try to avoid this simple but profound fact. They tell us stories about how fortunate we are that God blessed our parents and brought us into this world. They tell stories about free will and predetermination, reincarnation and fate. But what gets lost in these stories is the simple fact that we are not consulted about our existence. What is even more amazing is the fact that we are brought up to believe that this is a normal course of action and that we must accept our position in life, the luck of the draw, our lot. We are taught to respect our parents even when they don't deserve our respect; taught to endure whatever circumstances surround us and figure out a way to survive, grow up, and perhaps then acquire independence. By the time we become who we are, we are already custodians of someone else's views and ideas, someone else's bad habits and poor judgment. When things go well, we worry about the traditions of our aristocratic heritage. When things don't go well, we have the psychopathologies of our parents to deal with. By the time we are grown up we have internalized some of these worries. The rest of our lives, to continue this line of thought, are devoted to establishing identities independent of our heritage, the luck of our draw.

Psychologists are lost when it comes to explaining the fundamental injustice of our birth. Once we are born and raised, they have all too many explanations about anything having to do with our problems, our fears and fantasies, our inhibitions and expositions. But have they ever told us that it really isn't our fault that we are confused by the sheer incomprehensibility of the odds we deal with throughout life? Have they ever told us that some things are our problems and some we have nothing to do with? Have they made allowances for mismatches between who we are and where we happen to be born? The fact of our odd birth conditions, as I will call them, in relation to the person we wish to become, helps to explain why we are so confused and maladjusted as we begin our journey into adulthood. What can we do about this situation?

At some basic level, the answer is: nothing! We can't change evolutionary tides from sweeping across the human race. We can't prefigure or assign births that match a fetus with a set of parents. Regardless of modern medicine and the progress toward genetic engineering and test-tube babies, we are dealing with an unconscious existence, so asking a fetus to choose parents is somewhat out of the question. But once we appear on the scene, we should become aware of our limitations and the attitudes we must adopt to cope with a confusing state of affairs. In the twenty-first century we must also learn to deal with the technoscientific environment that surrounds us. In many ways, Western societies deal with a fairly safe and comfortable environment, one that provides some of our basic necessities, such as food and shelter. But it is also an environment that changes our wants, if not our needs.

I should hasten to say that what I just described in fairly gloomy terms might not be relevant to many of us. Not all of us feel that we were brought into this world under less than perfect conditions. Even those whose conditions are imperfect might not feel unwelcome or confused about the choices made by their parents in order to have them. Some might feel lucky to be alive, to be healthy and surrounded by the love of family. In those people a sense of responsibility and order is apparent from early childhood, a pastoral image documented so well by Norman Rockwell in his magazine covers and illustrations. But even for those who feel well adjusted, there is a gnawing feeling that life may turn out to be turbulent after all (when parents get divorced or lose their jobs during a recession). Even when your own life seems under control, there are enough stories in the media that indicate to you that whatever you're enjoying right now is temporary. Isn't this feeling and long experience in the United States what motivated the federal government to establish Social Security and other safety nets to catch us when we are about to fall? Isn't this background feeling also what has given rise to our determination regarding Yankee ingenuity and the drive to excel and improve our lots in life?

So, our own human condition and our own sense of how we are burdened with it are transformed over time, even though some of the basic tenets of the predicament of the existentialist view of the human condition remain constant. No wonder that we are doomed to remain frustrated! No wonder that our quest for a set of permanent criteria is problematic! Alternatively, no wonder we seek order as an expression of our increased responsibility to cope with technoscience!

Inevitable Anxiety

In the hyperconsumerist and technoscientific environment of America, we must cope with more variables than our ancestors had to some two or three hundred years ago. We find ourselves learning new skills every year, dealing with rapidly changing technologies and facing pressures from domestic and international competition without the ability to resist or transform these conditions of change. Our resistance, when it occurs, is registered as protest; our protest is interpreted as if it were social deviance and pathology. The cure, we are told, is to conform, go with the flow, be part of the success story of contemporary civilization. As we move along, we are expected to participate in predictable ways, so that we are rewarded with the knowledge that we too can survive the next stage of progress.

The human condition as understood by the existentialists of the past century was characterized in terms of the futility of human existence, the absurdity of our individual efforts to change the conditions of our experience, and a deep sense of our own insignificance in the affairs of the world. They, like Camus, underscored in their writings the absurdity of life itself. There is no rational exit strategy from the inscrutable maze in which we live. Whatever choices we make, they might turn out to be futile attempts to comprehend our existence and avoid our imminent death. Our lonely and absurd existence led some to propose suicide as a viable option to exit this world. It led others to propose focusing even more intensely on the individual rather than the social surrounding in which he or she lives, trying in vain to find some meaning. Yet, it led others, like Camus once again, to argue for hope, to believe that individual conviction and commitment can make a difference in one's life.

Perhaps the underlying theme of existentialists who suggested personal involvement in the world was how an individual deals with and responds to other individuals. Perhaps the goal was to establish a community of individuals who can interact in responsible ways and ensure the survival of the human race. Perhaps the method that was suggested dealt with coping mechanisms that ensured a sense of personal responsibility for one's actions no matter what results from them. Incidentally, the goal wasn't an idealized, utopian dream but a realization that the lonely individual might choose to commit suicide if she or he found no social involvement or commitment worth living for. Not much has changed in regard to the futility and absurdity of human existence. If gas chambers and atomic bombs (of World War II) are no longer the basis of the discussion of the human condition, then perhaps the inhumanity of bureaucratic life in advanced capitalism is—or environmental threats from pollution and overharvesting of natural resources. If the horrors of world wars are no longer front-page news in the Western Hemisphere, then perhaps the photographs and journalistic accounts of malnutrition and disease that devastate whole populations elsewhere in the world are. I'm not sure which is worse, the immediacy of the horrors of the past century for those who lived through them or the detached observation of current horrors through the media.

The former obviously inflicts pain on the individual experiencing the horror; the latter is sadly becoming a standard, alienated condition of human interaction. But what has remained similar in all of these different cases is the realization, recurrent as it haunts us, that despite our perceived comfort and wealth, we are still members of a civilization that inflicts pain on itself and cannot escape destruction and hate. Moreover, there is an increased awareness that we are all interconnected in this world, that disasters like the AIDS epidemic are not limited in their effects to one country or one continent. In the age of mass transportation and migration we are affected by anything that happens anywhere, and the effects are felt almost instantly. This global interconnectedness brings us back to basic human compassion rather than to lofty ideals about world peace, unless, of course, we insist on remaining detached observers as opposed to engaged participants. Rousseau's appeal to human pity as a noble characteristic comes to mind at this juncture. His insistence on paying attention to the way we view ourselves and others becomes even more pressing than ever before.

It's possible to argue that the title of this chapter is misleading, because there hasn't been a transformation of the human condition but only changes in the conditions of humanity. In the twenty-first century we are in the same position in relation to basic feelings about our mortality and the incomprehensiveness of the human fate as described by the existentialists during the past century. Perhaps the material conditions of our lives have changed for the better, but not our sense of anxiety and fear. We worry just as much as our parents and grandparents did. What has changed might be the things we worry about, but this doesn't mean that our worry is any less real to us than theirs was to them. Yes, they were worried for their lives when bombs fell in their neighborhood during World War II, especially in Europe. Yes, they worried about survival during the Great Depression in America. But we worry too. We worry about losing our jobs because a new technology has made our jobs obsolete. We worry about plane crashes and cancer from genetically engineered food and secondhand smoke. We worry about being left behind when new technologies sweep the landscape and present a picture of the world to us in a format we have never seen before. So, the transformation is of the conditions of humanity rather than the human condition as such.

We remain vulnerable as ever before. No matter how fortunate we feel in the midst of technoscientific wealth, with comfortable cars and stereos, with hospitals and central heating, we are still as lonely as ever in contemplating our very existence. One need only observe the popularity of Web sites for singles and the chat rooms that spring up daily on the Internet to appreciate the ongoing desire to meet other people and connect to them in some meaningful fashion. We seek approval and love, the support of our fellow humans as family members and friends, and we know that we might be disappointed.

We know deep down that we can be let down by those closest to us, by those whose support we'd like to take for granted. We know we have to fight for our personal peace of mind and our happiness, just as much as for definitions of who we are and solid identities for ourselves. We know that promises can be broken and that hopes can be dashed. When everything is said and done, we can't change the world to suit our needs and desires, our tastes and dreams. We have to adjust and learn to cope with our surroundings, agree with others when we'd rather do things our way. Our loneliness can't be swept under the rug, because it stares at us daily. The absurdity and futility of our existence haunts us too. The demand for a responsible personal existence has not diminished under the new conditions of the twenty-first century but has been amplified.

Technoscience seems to have done very little to mitigate these conditions of horror and pain, hate and destruction. Technoscience might help create barriers to the immediacy of the experience, as in the personal pockets of comfort we create for ourselves and the transmission of data through a media that filters out some of the more devastating occurrences in the world. At the same time, our pockets of comfort seem less stable and reassuring over time. The classical British notion of one's house as one's castle seems less convincing nowadays when the American dream of owning one's home can turn into the American nightmare when three mortgage payments are not paid. Likewise, the transmission of data through the media and the Internet is more invasive than ever before, making our reception almost inevitable, unavoidable. We cannot ignore the strife of faraway people just because they are geographically remote. These people end up being presented to us in our homes through print media and television broadcasts. In order to better understand not only how but why we have allowed ourselves to be transformed in the ways we have during the past century, it might be useful to examine some of the myths we have cherished and carried with us over the years. These are myths that allowed us to let down some of our (psychological) protective mechanisms (those that insured an insulated view of the world) and made us more open and vulnerable, more emotionally exposed and fragile.

My purpose in examining some of these cultural myths is to explain the cultural and psychological conditions under which we live in contemporary (Western) society. These conditions necessitate a stronger commitment to live a responsible life—that is, take responsibility for ideas and actions that seem remote from us. Herein lies a plea in every person's life for an ethical dimension that cannot be summarized or dismissed in religious terms. We can no longer claim that just because we go to church on Sunday and make some contributions to our church, we are being ethical. We can no longer claim that just because we donate some money to the endangered spotted owl fund, we have done what we should. Rather, our personal responsibility goes deeper and wider than that. We have to watch for how we might inadvertently impose our values on others, how we might violate someone else's privacy when we use the Internet, how we interact in the workplace or the university system with others, and how we conduct ourselves in public. Our personal lives have become part of the public domain more than ever before.

Any of our actions has the potential to affect others in a manner never conceived of before. Therefore, an increased sense of personal responsibility and a heightened sense of humility and dignity should be in place in this twenty-first century.

Do my injunctions and admonitions sound like a sermon? Perhaps. When they do, I hope they remain in the tradition of the biblical prophets who found it their responsibility to alert their audiences to the dangers that lie ahead.

CHAPTER THREE

The Myth of Freedom in the Information Age

Knowledge Control

I finished the last chapter with a plea regarding the personal responsibility we must take for our actions and their effects on other people, as we get a better grasp of the conditions that brought us into this situation. This notion stems from the belief in the ethical dimension of our existence. Through the study and examination of the history of ideas, we can glean two major views that surface pertaining to human behavior. Put in basic terms, they can be identified as dealing more prominently with either the heart or the mind. The one that deals with the heart suggests that humans are ruled by their emotions. It is our emotions that color our vision, that determine what we learn to accept or reject about ourselves and others, and ultimately about the world at large. Some people, whether artists or spiritual poets, associate this view with a romantic view of the world or with an inspirational outlook on life. According to this view, in order to change our behavior we must affect our feelings. For example, if we are attracted to another person, we will be attracted by that person's ideas and concerns. If a teacher has charisma, this teacher can make mathematics as well as art seem important and worth learning. If a political leader has charisma, if we are mesmerized by the way she or he appears on the television screen, we are more apt to identify with the ideas that she or he expresses. Education and business models have been created based on this view of human behavior in order to explain to teachers and leaders how to motivate their students and employees, how to appeal to them.

Incidentally, psychotherapists of one kind (e.g., Freudians of the early part of the twentieth century) believe in the first view about how feelings determine our behavior when they suggest behavior modification through punishment and rewards. When pain is associated with "bad" or inappropriate behavior given the standards of the day, then the presumption is that the emotional (and physical) rewards of instant gratification will lead us to understand how to behave. Humans, just like mice in experimental situations, don't like electric shocks and will avoid them in future instances. The mind learns from the body, and the body is perceived as a mass of nerve endings that determine our emotional and physiological states. Perhaps this is also related to the mind-body problem, discussed hundreds of years ago by the empiricists and the rationalists. But instead of rehearsing their ideas and debates, suffice it for us to appreciate here the relationship between the mind and body, between the mind and the heart.

The other view of people or of human nature focuses more on the cognitive abilities and features of humans as rational thinking animals, a view dating back to the Greek thinkers, such as Plato and Aristotle. It suggests that the mind has the power to control the emotions. We can think through an issue and then be attracted emotionally to the idea or the person who professes it. Some of the proponents of this view have been associated with the Enlightenment movement of the eighteenth century, not to mention all the philosophers of previous periods. This also means that knowledge takes precedence over intuition or gut feelings. For example, even though the pain of studying or doing hard labor might disgust us, we might be convinced that it's worthwhile to do either thing. We might listen to reason and be convinced that years of learning will pay off at the end and will eventually lead us to be happier with our lives. We might be convinced, rationally, that love at first sight isn't the way of finding a soul mate with whom we'd want to spend the rest of our lives. Instead we should weigh the characteristics and attributes (and shortcomings) of the other person before committing ourselves to falling in love.

I begin this chapter with this brief introduction in order to explain its structure and the method by which it proceeds. I prefer to hold onto to the cognitive view of humans and believe that the mind can control our emotional lives, and that the mind has the power to make us feel happy or sad. The acceptance of this view, even if only provisional, is essential in order to accept what this book tries to illustrate. For it insists on tracing the history of ideas, finding the theoretical makeup of our worldviews and the ways in which they have influenced our cultural formation in the Western world. Moreover, the acceptance or rejection of this view becomes integral to the acceptance and rejection of the proposals for personal behavior and taking on responsibility as laid out in later chapters. I should hasten to add that this is not to say that we don't have emotional outbursts or that at times we are not overcome by inexplicable feelings. But this is more the exception than the rule in the lives of most people. Even when one experiences an emotional outburst, it can be explained rationally. That is, we can analyze the situation in rational terms and find the variables and conditions that led to the outburst. Oh, you may say, this is hindsight rationalization; this really doesn't help much. Perhaps not, if it's indeed after the fact. But what if you explain the situation in which you live ahead of time and are then able to judge situations as they come along, have the ability to explain to yourself why you feel the way you do? Wouldn't this be a contribution to a more harmonious functioning of a society?

This, to me, is one of the most important keys to leading a successful life (at least in the sense of understanding yourself and respecting the emotional turmoil you might be experiencing). In other words, the cognitive approach presupposes an appeal to some ethical or behavioral code. If the notion of a presupposition strikes us as too vague or unhelpful, we might choose to realize that a cognitive approach necessitates or leads to the development of an ethical code of sorts, a set of criteria by which to evaluate situations and make decisions. The code itself could be handed down through religious or political frameworks, through myths or narratives, folk tales or tradition. In either case, there is a standard or set of standards one can assess (accept or reject) and revise in terms of new circumstances where the standards no longer apply or new ones need to be invented.

If you accept even some of what I've just said, then it would make sense to you why I try and explain the cultural and ideological conditions under which we live. You might get a different perspective of what preceded your own situation and why certain elements of your life are not within your control (and therefore you shouldn't beat yourself up when they happen to you). It might help you see where it is that you can make a difference, what changes you can bring about, and how important it still is for you to conduct yourself in an ethical manner. By "ethical" I mean how you understand the rights and duties you have as a human being and as a member of society, the way you fully accept responsibility for your behavior. Finally, by ethical I mean the process by which you learn to accept and reject the rules and regulations by which a community lives. This is important so that when you construct your own rules of conduct you can justify them not only to yourself but also to those who might be affected by them.

If everything so far seems a bit abstract, let me give an example from the life of the academic world, a world in which I have lived for over twenty years. You can look at fellow students as unfriendly competitors who must be beaten at the intellectual game, or as fellow travelers from whom you can learn something. You can master the system to get good grades or a promotion without really doing the required work or really learning anything. The system itself provides only guidelines and some rules. You have the power to interpret them in a useful and engaging way. You can look forward to fulfilling a research assignment and "get into it," or you can do the bare minimum to receive a passing grade. You can view every assignment as a challenge and an opportunity to change your life or as a burden to avoid ("cutting corners" as skillfully as possible). It's up to you what you make of your situation.

You might be shaking your head and saying to yourself that this is all fine in theory, but in the real world it doesn't work. You might be thinking about the material differences between rich and poor students, between those who must work to pay tuition and those who don't. I don't disagree that there are real circumstances that make a great deal of difference in how we function in this world, and that the academic world is no different from the workplace. Poor people with great ideas and ambition are less likely to get bank loans than their rich counterparts. Poor students are less likely to have the time to read through assignments (because of work and fatigue) than rich ones. If we stopped our examination here, we would limit our discussion to the material conditions that provide obstacles to the poor and less prepared among us and perhaps make them so disheartened so as to give up on making any changes at all.

Instead of pretending that material conditions don't matter or that they can be changed overnight, I suggest that in order to change the material conditions, we might want to focus on changing our minds and hearts, our attitudes and outlooks. Perhaps if we understood better how things in the world have been taking shape over the past couple of centuries, we would find many keys with which to open new doors, new opportunities. No, I'm not advocating that we accept inequality and the injustices we observe around us. Rather, I'm advocating the pursuit of knowledge and understanding—cognitive tools and methods of engagement—as the first step toward change. That first thinking step is both outward looking (toward the culture at large) and inward looking (toward the psychological conditioning we have endured in our lifetime). Perhaps this is why the first step is so hard for so many to take; perhaps this is what turns them away from a book like this; and perhaps this is what makes them complacent and detached from the affairs of the state. Reflective thinking, critical analysis, and cognitive agility are not easily undertaken in a culture that feeds us sound bites and prefabricated ideas. My role as an educator, then, is to translate pieces of knowledge into operational methods by which people can understand their cultural surroundings and accordingly form their responses and actions.

American Myths

I offer here a little cognitive tease regarding the cultural development of America since its inception, since the days it broke the monarchical yoke of the British Empire. Just as we discussed earlier the shift in worldviews from the notion of abundance to scarcity within the framework of the biblical narrative of the Garden of Eden and the subsequent fall of Adam and Eve from grace, I wish to examine a cultural narrative embedded in the American mind. There is a little secret that takes a long time to discover: freedom is a myth. We are constrained in so many ways that the idea of freedom might seem like a cruel joke someone is playing on us. When we seriously consider the idea of freedom, we would have to conclude that we are not completely free to do whatever we want or be whoever we want to be. We are born with certain features and specific genetic composition. No matter how hard we try, some of our features cannot be changed. Then there are mental and emotional constraints that determine how we think and feel and how we react to whatever happens to us. These constraints also determine how we behave in social settings and how socially adjusted we are among family members and friends. So, what does it really mean when we say that we are free to be whomever we dream to be? How does this idea play out in the technoscientific age, where the freedoms we seem to be afforded are constrained in many ways?

Rousseau warned us about being born free and finding ourselves chained by social constraints. What he meant was that there are many unwritten rules we must obey, as well as many laws to which we must adhere. The late-eighteenth-century German philosopher Immanuel Kant too was concerned with our lack of determination to free our minds from all the nonsense that surrounds us. For him, though, it was more of an intellectual obstacle than a legal one, for he worried about how free we are to engage new ideas and put aside all superstitions that have traditionally clouded our thinking. The whole Enlightenment project of the eighteenth century was one big effort to free people from the dogmas of religious doctrines and the oppressive powers of kings and queens. Either the church or the state was the enemy; only the individual could free herself or himself from them all. This wasn't an easy task, because it required shedding the old and embracing the new. One had to be willing to take the risk of incorporating something novel without the benefit of years of experience.

The church had years of tradition behind it when it proposed ideas and sanctioned certain modes of behavior. It knew what would work well and what would be dangerous. It knew, just like kings and queens, under what conditions people could be scared into believing what they said and doing what they recommended. But this would all evaporate if the individual became the master of her or his own destiny. The risk was that a wrong choice would be made and terrible consequences would ensue. Would promiscuity lead to free thinking, a more responsible outlook on one's behavior? Or would it rather lead to general confusion and the denigration of human dignity and self-respect? These are questions that were raised a few hundred years ago but that recurred during the civil rights movement of the 1960s (when people of color and women gained power). Would the introduction of contraception lead to the protection of women's lives and rights or unleash irresponsible sexual behavior? The balancing act was one of taking risks and making them reasonable and calculated so as to improve the human condition, which was the goal.

How would one accomplish this goal? How could an entire society ever accomplish this feat? As far as the Enlightenment leaders were concerned, it was both psychologically difficult and mentally (cognitively) easy. All one needed was will power and brains and they could turn their lives! In other words, one could become the king or queen or pope of one's own life. A person's entire worldview could be altered by the way she or he interacted with the world. Literacy rates in the Western world increased dramatically, and more and more people got some form of education. Reading texts that had never been available to the masses (of course, the introduction of the printing press helped make copies of books available at low cost) opened the way to personal introspection by more people than ever before. For example, by the mid-nineteenth century the United States had opened its treasury to establish land-grant institutions of higher education. The public was invited to participate in the acquisition and use of knowledge, and was able to take personal advantage of what had been available until then to only the few who were fortunate (and rich enough). Education became mandatory in the Western world by the twentieth century, and most citizens were expected to have matriculated from high school and pursued some level of postsecondary education. Moreover, the technoscientific age ushered new devices by which to make public education more accessible and cheaper to greater number of people in more remote geographical locations than ever before. Our Western culture endorsed the implications of the myth of freedom with open arms: greater education and little resistance to public funding of skill acquisition. This was a dream no one could refuse to accept. This seemed too good to be true—and indeed it was!

Perhaps one critical element that needs mentioning here is that just as there were obstacles to the education of the masses in the Enlightenment period, so there are obstacles in the technoscientific age. The sticking issue can be reduced to the availability of money to pay for education, to disseminate information across social and economic barriers, and to provide similar quality of education everywhere to everyone. Perhaps the demographic composition of contemporary Western culture differs from the one of two centuries ago, but hierarchies of power relations and economic inequities remain intact. The nebulous and ever-changing group of rich citizens suffers less and enjoys more of the fruits of technoscience in comparison to the ever-increasing group of poor people. This fact hasn't changed over time despite the changes in the apparatus through which education is provided.

The information age illustrates the urgency regarding freedom. The freedom to be left alone differs from having no constraints to follow your heart's desire. In the former case we think of limiting government intrusion into our daily lives; in the latter we think of our own pursuits and ambitions. These two interpretations come closely together in today's environment. Computer technology promised to democratize the Western world and provide information technology to everyone relatively cheaply. Just as calculators became extremely cheap and available to anyone who needed them, so the belief spread that by the end of the twentieth century everyone would own a personal computer and have direct access to the information "highway" through the Internet. Have these dreams been fulfilled? Not quite. There are still discrepancies between the rich and poor, and they are openly expressed through the ownership and use of the most recent computer technologies. Databases are more accessible to some than to others; knowledge of how to use the technology is reserved for the better educated and better informed among us. Finally, the freedom enjoyed by the mastery of these technologies has turned out to be elusive as well. The cell phone we carry with us keeps us at work even when we leave the office. The laptop computer we sling over our shoulders ensures our connection with and response to our employers or consumers wherever we happen to be. Finally, the information age has given us information about the world that surrounds us, but it has also given information to others about where we are, what we do, and with whom we interact. It allows others to monitor our personal behavior and consumption patterns. In short, the information age is also a super-age of surveillance; we are informed as much as we are being informed on by others. On some level we recognize this new predicament of our modern existence, but we refuse to resist the encroachment of technoscience, embracing every new device as if it could transform our lives. Why?

The Price of Freedom

Despite all of these examples, I would hasten to add that the myth of freedom is too seductive to resist and so are what we take to be its instruments of application. The endorsement of the myth and its attendant technologies offers a way to rationalize and justify social and economic rewards. These rewards help improve our position in society and entice us with a greater sense of freedom, false as it may be. For example, people are induced to "think outside the box," be creative and innovative, aspire to intellectual and artistic levels they would have thought weren't possible. People are encouraged to take chances in life, be adventurous, even befriend other people who traditionally wouldn't be proper to associate with. What's the price? What do we have to give up in order to "gain" freedom? These questions are not silly philosophical questions but rather practical questions that are or should be in fact raised by us at some point in our lives. When students drop out of school, are they merely pursuing their own freedom of choice and a sense of Freud's "pleasure principle," or are they considering their pursuit of freedom? If we don't openly discuss these questions, we might end up endorsing an idea that is impossible to fulfill. If we don't ask these questions, evaluate our situation cognitively, we risk losing more and more of our freedoms. Moreover, we might get frustrated with our culture, even angry at ourselves, because we misunderstand the complex conditions under which the myth of freedom has been politically informative in the formation of the American state. Here, too, my plea for knowing and understanding what freedom means in the abstract and in practical terms ends up affecting our emotional and psychological state, and it allows us to control our responses to the application of this myth in responsible ways.

Like many myths, the myth of freedom is powerful. Indeed, the promise of absolute freedom to pursue any dream is powerful enough to convince young people to leave their families and go to war to become heroes, not to mention leave the state to earn more money in strange places and even foreign soil. We are willing to do quite a lot to gain freedom, or at least embark on the path that presumably leads the way to gaining freedom. We are willing to take risks and undertake personal hardship for long periods of time for the promise of freedom. Not unlike Jacob in his labor of love for Lavan (to have his daughter's hand in marriage), we are willing to pay a high price now for future rewards. Moreover, if we are able to couch our present situation of some hardship or heartache in terms of future liberation, then we can endure for longer periods and with a greater sense of accomplishment than if we couldn't. The American dream, dating back to our founding fathers, has always played off this myth in the political arena—whether in terms of the British crown or in terms of entrepreneurship, all the way into the computer age. We fought for freedom against a colonial rule that imposed taxes without representation, and we still fight our own government agencies whose regulations impede our freedom of commerce.

For example, many people immigrated to the United States over the years in order to find the freedom they could not obtain in their homelands; many came to pursue their own dreams in a place that would allow more opportunities for upward mobility than they had ever known. They sold all their belongings to afford the voyage across stormy seas. Many still risk their lives today under precarious conditions (unsafe small fishing boats without drinking water or across the southern border with Mexico in converted trucks without breathing room). For the most part, the promise has been fulfilled, whether on the shores of New York or on the western trails to California. More individual rights and freedoms are observed and legally protected in this country than in many others. But there is a caveat in regard to these freedoms that must be revealed when we talk about freedom. Every one must pay a price. As I asked earlier: What is the price? Is the price itself a hindrance to the acquisition of freedom?

These are tricky questions, for the price we pay is different in each case. For some of us it means paying taxes to a government that is supposed to protect our rights and freedoms with a military and a legal system. For others, this means having to work for a corporation whose policies are strict and binding and even oppressive. Still for others, it might mean having to abide by a social contract of sorts that dictates behavior in accordance with certain social conventions and moral norms, such as a dress code or the ownership of certain material conveniences. For still others, it may mean compromised privacy or being under electronic surveillance all day long (and even after working hours). Is the price too high? Perhaps the price is just right if we learn to redefine what we mean by freedom. If freedom is not absolute freedom but only a qualified freedom, as is the case of college professors who comply with academic codes of conduct (how to interact with students, how to relate to colleagues and administrators) so as to retain their academic freedom, then the price might be just right. If freedom is understood to be inherently limited to those actions that we undertake in isolation so that no one else is affected, then perhaps there is no problem.

The British philosopher and political theorist Isaiah Berlin commented on the positive and negative aspects of freedom, explaining the difference between having the freedom to do anything you want and having the freedom not to be bothered by others (1969). Both forms of freedom help explain how we must balance our own personal freedom with the personal freedom of others in the same society. This means being able to count on certain protections from the intrusion of others, from the abuses of others who are more powerful (or just louder), and having set boundaries that everyone must honor. Some might even liken this discussion to the Golden Rule, which says that we shouldn't do unto others what we don't want them to do to us. Kant's moral philosophy is based on this simple but profound principle, insofar as he insists that each one of us examine her or his behavior in terms that can be universalized to everyone else in a society. For example, you shouldn't lie to others, not only because it's not nice or mean but because you don't want to be lied to by others. Similarly, you shouldn't steal from your neighbor or employer, because you don't want to give in this way permission or license to steal from you. This also means behaving ethically without needing to follow the dictates of a particular religion. You see, Kant is arguing for the implementation of this principle almost on self-interest grounds and not on an appeal to a lofty ideal or a wrathful God.

Since freedom is defined socially and culturally, politically and economically, it is applied in every aspect of our lives. We can think of financial freedom as being rich enough not to worry about our daily expenses. We can think of social freedom as being eccentric and disregarding some conventions relating to our clothes or our eating habits. We can think of political freedom in terms of free speech and the formation of political parties or lobbying groups. However we exercise our personal freedom in any of these domains, freedom is not so much a cultural myth that is couched in political terms and removed from us but a lived reality we interpret for ourselves with the help of our peers. There is a social context within which we exercise our freedom, knowing that our behavior extends beyond ourselves and realizing the (legal and social) boundaries that protect us from intrusions by others and the government. To be free is to acknowledge the constraints imposed by others. To be fully free is to realize that these constraints are essential to the maintenance of freedom. To be content with this realization is to be socially adjusted and psychologically mature. Finally, to be free within a community is to understand one's responsibility to the entire community and the environment as well.

The Pleasure and Reality Principles

But wouldn't we then revert to the unfortunate position described by Sigmund Freud in terms of the Pleasure Principle and the Reality Principle? These two principles were for him the driving forces behind human behavior and the way society would or wouldn't sanction it. Wouldn't our desire to fulfill pleasures be restricted by the harsh reality of social dictates? Wouldn't we be frustrated rather than content when we realized that we had no freedom at all except to follow the rules and regulations set up by someone else, namely, satisfy our desires within the parameters set by others and according to the tastes and preferences of others? Where is our own power to make up rules? Where is our quest for power, our "will to power," as the nineteenth-century German philosopher Friedrich Nietzsche calls it (1967), expressed or legitimated? If all we can do is follow someone else's ideas of right and wrong, how free are we really? Can we even intelligently talk about freedom anymore?

These questions bother us in one way or another. They don't only bother us when a police officer stops us for speeding or for not wearing a seat belt. These incidents only exemplify what some of us normally think about and feel deep in our hearts. These questions don't only surface when we receive our paycheck and see how many taxes are withheld. They don't only appear when we need a permit to renovate our home, buy a car, or enroll in an electronic mail service. Rather, these questions are always in the back of our minds; they keep festering in us, and the longer we live the more they appear as part of our normal lives. In short, these questions turn into persistent elements in our existence, habits we cannot kick, so that they are taken for granted after a while and become unnoticeable.

Perhaps we can think about freedom in this context in more realistic terms than those provided by philosophers or political theorists. Perhaps the conceptual framework of freedom has altered over time or needs to be revised in light of the experiences we have had in America in the past two hundred years. Incidentally, this shift in thinking and theorizing is similar to the one discussed in the previous chapter in relation to technoscience—the mutual interaction between the theoretical domains of the scientific enterprise and the technical and engineering developments that inform them. As political, economic, and personal freedoms have been challenged and encoded in our culture, they have necessitated the transformation of our theories and models (and made explicit those principles we took for granted implicitly). Perhaps we can think about freedom in this context as the personal control we use to negotiate the boundaries between ourselves and the community in which we live. As I said earlier, the lines of demarcation between the public and the private domains have been blurred over the years, partially with the intrusion of technological instruments, partially with hypercommunication, and partially with the growth of government agencies. Because these lines have been blurred over the years, the question of personal freedom has been reintroduced not only in vague political terms but also in terms of individual rights. But how much control over the intrusion of the public into our personal lives do we have?

Things we'd like to control can be divided into two kinds, as the ancient Greek Epictetus reminds us: the ones we can do something about and the ones we can't. There are things in life that come within the range of our control, which allow us to interfere, so as to change the course of our lives. But then there are those factors, conditions, and situations in life that are completely beyond our control. This delineation is clear and simple: we cannot change the weather, but we can change our disposition about the weather (taking an umbrella along or enjoying the beauty of white snow). We cannot change our height, but we can change our attitude about it, and so on. Knowing which things belong in which class can protect people from frustration and anguish. We become frustrated when we confuse what is within our control with what isn't. We get overly concerned with what people will think about us, as if their thoughts are within our control. What is within our control, though, is how we react to what they think or say. In other words, the key is becoming aware of what is in our domain of control. We have the power to ignore them, minimize the effect of their words on us, agree or disagree with them, and if necessary, take their criticism as helpful comments about how to change something about ourselves. We are better served (in terms of personal freedom and healthy mind-set) by exerting our energy on those facets of our lives over which we do have control.

One could argue that these stoic ideas were presupposed by the Enlightenment leaders as they embarked on methods of reasoning and rational evaluation of ideas and practices. Since they believed in the cognitive view of human behavior (and the control of human behavior), it stands to reason that rational discourse and critical analysis are the appropriate tools with which to approach decisions in life. The criteria themselves are set by humans (and therefore are within their domain of control), and the decisions made in light of them are also made by humans (and therefore within their domain of control). Furthermore, since the criteria and decisions are within human control, any changes taken when errors are detected or when negative consequences appear are within the control of the same people. They need not appeal to any force outside themselves, nor do they need to seek the permission of monarchs or priests—they can change their minds on their own, be free to take risks and regret or enjoy them, as the case may be.

Once we appreciate the power we have and the control we can exert on various parts of our lives, we might be less worried about what others do and focus more on how we engage with our world. But to do this effectively, we must exercise a virtue that was emphasized by the ancient Greek culture and emulated by European thinkers and cultures in different forms over the past three hundred years. When we effectively exert our control, can we indeed feel responsible and free, feel that we are indeed independent from the world while recognizing its constraints. For example, we can make certain choices about our lifestyles, about the kind of careers we'd like to pursue, even where we'd like to live. We can decide to befriend certain people and not others, buy into certain temptations and avoid others. It's within our control not to be drunk, for example, if we don't have a predisposition to be alcoholics. We can choose to work in the public or the private sector, fix cars or computers, develop new software or hardware. If we love flowers and trees, we can become landscape artists. If we love animals, we can work in an animal shelter or an equestrian center. In short, we have many options of what to pursue if we listen to our own sense of enjoyment and fulfillment.

This is not to say that we can survive without food or beverage, or that we don't need a whole community to help us live the kind of lives we enjoy. Neither is this saying that "the sky is the limit," that anything is possible. Of course, there are limitations on our choices and the skill required for some of them. Yet, what I'm suggesting here is that we can accept some of the natural limits and social codes and not worry about changing all of them. Each one of us can navigate and negotiate some of the terms of her or his existence so that the choices out there don't seem overwhelming, daunting, or unforgiving. We can accept that some things in our lives are "givens," being what they are. Acceptance, from my perspective, is not resignation or complacency but rather a realization that we should fight only some battles, only some of the time, as opposed to continuously fighting every situation as if our lives were numerous battles within one enormous war.

In order to be effective and responsible in the ways we exert our control and express our freedom, we must be prudent. Prudence, as the ancient Greeks taught us, is a way of accounting for justice and moral norms within a set of conditions that are part of our lives. This accounting, incidentally, is a method of coping responsibly with the situations in which we find ourselves as well as the our own choices regarding these situations. Prudent compromises are necessary for emotional stability, easing the tensions we might have. We might not fulfill lofty goals or be as moral or as free as we'd like to be, but we can avoid hurting others, maintain our dignity, and interact responsibly within our community. The proposal to follow the advice of the ancients, then, is a proposal to live more comfortably with ourselves and with others. This is a way of reducing frustration or stress and of achieving some level of peace of mind, as Spinoza recommended some four hundred years ago.

Looking back at some aspects of ancient Greek culture in conjunction with how to lead our lives is also a way to respond to the predicament that Freud outlined. For Freud there was no way out of the frustration of individual desires. The pleasure principle—those things that would presumably make us happy, such as unrestrained sex—would always be curtailed by the reality principle—it is impossible to run a society in which everyone's sexual desires are always fulfilled. This predicament can express itself as neuroses or as psychoses—namely, in the forms of personal mental illness. We are prone to feel frustrated and angry at a world that doesn't allow us to do anything we want whenever we want to do it. We are prone to get so frustrated and angry that we feel upset and paralyzed, feel that it's not worth fighting for our goals and ideals anymore. We might end up feeling, like some disillusioned teenagers or middle-aged workers, that there is no way to ever get beyond the conditions in which we find ourselves. Once this feeling takes over our existence, no argument in the world, no motivational guru, could ever change how we feel about ourselves. So, instead of drinking ourselves into oblivion, my own recommendation is to fall back on an appeal to our mind, our thinking and our reasoning, the way in which we have distinguished ourselves from the rest of the animal kingdom. This is why I offer the present way of perceiving our lives, critically evaluating what we can do to improve our lot in life and resolve the predicament into which we were born.

Unlike the Freudians of the twentieth century who analyzed the predicament of the human condition and offered clinical therapies for individuals, the Greeks tried to dissolve the predicament by resolving the conditions that caused it. As mentioned before, they argued that some things are not within your control, so don't fight to control them, sex included. For instance, you cannot "make" someone fall in love with you, but you can fall in love with someone. If the one you desire doesn't desire your company, find out if it's within your control to change her or his mind; if it is, do something about it; if it's not, let it go and move on with your life. Perhaps the discussion of control is about letting go of those parts of your life that you cannot change and learning how to cope with those aspects of your life you don't particularly like. If you have only limited energy to devote to yourself, invest it well. Perhaps what is advocated here is cognitive therapy: thinking critically through a problem, finding alternative solutions, and choosing rationally the best coping mechanism (alternative solution) according to set criteria (however defined and outlined). In case of success, repeat the procedure. In case of failure, reassess the procedure or the set criteria. Remember, if at first you don't succeed, try, try again . . .

So, from freedom we move to control, and from control we move to coping methods that consider energy and time management. The sense of self we develop in the process of maturation depends on a sense of freedom. Freedom for us isn't an ideal or a noble concept but a way to think through and assert our identity and independence. Whether we insist on referring to freedom or not, we do try, as hard as we can, to establish and maintain our autonomy (so as to establish our sense of identity). To be autonomous, then, is to have a sense of freedom that allows us to exercise our desires as best as possible. Kids insist on and fight for their independence as they grow up. They want to find out who they are, despite or because of their upbringing. They insist on being themselves, setting up their own framework for expression and growth. Professionals work hard to be financially independent, to set themselves up as entities who define their own fate. Retired people want to remain independent from their children or the hospitals and nursing homes into which they might be condemned. Just as kids don't want to be known as the daughter or son of someone, so grandparents don't want to be known in terms of their children and grandchildren. This is why they insist on telling us stories about their glorious past, about how they made their way in the world, about who they were and still feel they deserve to be. Everyone desires a sense of control, because they fear insecurity and the loss of autonomy.

The Stoic Sense of Control

Individual independence in contemporary culture is primarily defined in financial terms, even though its roots are more fundamental and can be detected regardless of economic conditions. Financial independence, then, is only one (however pronounced in contemporary capitalist society) expression of what we think we need to have to assert our individuality and our personality. How much money you have in the bank defines who you are on the social ladder and the respect (or disrespect) with which other people treat you. All of these expressions of wealth or poverty amount to a picture you paint of yourself, even though the process of painting the picture is highly influenced and motivated by the way others perceive you. Yet, in the process of constructing your material being, you are also constructing your inner being, whether you want to admit it to yourself or not. As you keep on refining and modifying this picture, you are also experiencing a powerful sense of your identity. This, perhaps, is where you encounter doubts about how much freedom you have to paint your picture the way you want it painted as opposed to how it's supposed to be painted in light of social pressures.

The reason for discussing the notion of independence and its relation not only to the constitutional framework of American society but also the economic context in which it finds expression, our personal lives, has more to do with the boundaries of autonomy than with the limitation of freedom as such. In other words, what is at issue are boundaries, limits to our freedom (whether dictated by others or internalized and dictated by us). These limits are real and are met by us every day. They are not conceptual constructs or abstract ideas that philosophers argue about. Therefore, I'm more interested in explaining the difficulties of delineating the limits and the order of personal freedom than with the psychological and emotional determinants of personal autonomy. This means that at the heart of the struggle to be free, to be all you can be, as the U.S. Army commercial says, is a struggle to define parameters, boundaries, and a framework. In short, there is a cognitive (mental and intellectual) struggle to maintain one's sense of freedom within a fairly disciplined life. How much order is enough? How much discipline is too much? Can we develop an open-ended order? Once these questions are raised, it makes more sense to begin thinking about the very personal as well as the social need for order as an instrument and basis of responsibility. Order here could mean rules and regulations that set the tone for what we should and should not do, and how our actions affect other people. It could also mean the simple habits we adopt over the years and stick by in order to make sense of our surroundings (coffee in the morning, eight hours of work, beer on the way home, dinner and television, bed and sleep, and the cycle repeats itself daily). Once again I return to our mind set or attitude in regards to the myth of freedom as a way to understand what affects (frustrates, reassures, and sets) the parameters for our choices in life.

Rousseau was of the opinion that the only true freedom must be sanctioned by society as a whole. In order to accomplish this, he suggested that we enter into a social contract that sets some limits to what we can and cannot do. Once we all agree, on some general level, that we are all part of the same group and that what applies to me must apply to everyone else (the so-called Golden Rule), it would be reasonable and liberating to exercise our freedoms. Freedom is then sanctioned and protected by the entire society; it is responsible behavior in the face of uncertainty. It is not the individual anymore who tries to assert a personal desire but an individual who is understood to be part of a larger community that knows how best to fulfill a need or desire. It seems that traumatic situations in the life of a nation, such as wars or natural disasters, illustrate more explicitly these issues. For example, the behavior of individuals under these circumstances are recognized simultaneously as personal actions and as the actions of members of a society. But as individuals act alone, they act on behalf of a nation and thereby symbolize the mind-set and character of a nation.

Cultural signposts orient our choices and delineate the boundaries of our actions. They determine—through custom, habit, tradition, and their respective expressions in the media—how we should behave. When they are critical, they even explain why we should behave in a particular way. One could think in this context of the different personal reactions during the draft in the Vietnam War and World War II in light of the cultural messages that were delivered. Patriotism in one case and public protests in the other belie the resolve or confusion that underlined American policies and the moral debates of the day.

Being part of a community (including family) might be understood by some as a loss of individuality or even a way of conforming to the group. But even within the family each individual is treated differently and is allowed to express a personality—she is creative, he is bashful; she is assertive, he is introspective. Each person falls naturally into a distinct role, but not everyone is allowed to freely enjoy playing this role. We expect to find differences among people as we get to know them better. This is no different from a doctor who asks for family history in relation to illnesses and medical conditions, knowing all along that the particular medical history of the patient is unique. Knowing the medical conditions of family members is only one step in diagnosing the patient as an individual. Knowing the rules of a society is likewise only a first step toward getting to know a person who belongs to that society. Likewise, knowing the rules allows the individual to navigate and cope with these rules on a personal level, with variations and adjustments. Coping mechanisms are ways by which people assert themselves, so that they don't get lost in the great social shuffle. Loss of individuality is tantamount, in our view, to loss of freedom. But this need not be the case, since on some fundamental level we never lose our individuality or our freedom.

To keep our individuality within and beyond the community, we need to maintain a certain balance between attributing too much of our characteristics to the group—sometimes seen as group stereotyping—or too little, believing that we have nothing to do with the group. This balance lets us know how much we owe to our ancestors and how much distance from them we can claim for ourselves. It is this sense of balance that keeps us looking forward and backward at once, keeping us alive without stifling our progress. In short, this is a balance that respects the boundaries between the self and the community, between being part of a group and apart from it. For me this means neither pride nor shame in my roots and background, but a critical acknowledgment that I am a Jew who was born in Israel to refugees from Germany just prior to World War II. It also means that my academic credentials (as a professor of philosophy) and affiliation (with the University of Colorado at Colorado Springs) are part of what I do but that they do not define completely who I am. I am also a son and a sibling, a father and an uncle, a developer and a restaurateur, an art collector, gallery director, and author. Which part should I isolate? Which part should I foreground? Which part is "representative" of who I am? Am I free in all parts of my life in a similar way? Obviously not!

But as I cope with my own situation, I learn to express those parts of my personality that I think are more important than others. This means, for example, that I must earn enough money to support not only myself and my daughters but also a lifestyle commensurate with certain aspirations they and I have. These aspirations, I am well aware, are imposed on me by the American culture in which I live and the social circles (of academics and business people) to which I belong. There are certain expectations from a college professor (book collection, extensive travel, cosmopolitan experiences); certain expectations from those living in the West (all-wheel-drive sports utility vehicles); and certain expectations from business associates (belonging to organizations, donating to charitable funds). Yet, I'm able to draw my own lines of consumption, accept and reject invitations and solicitations, agree or refuse to participate in certain outdoors activities, like hiking, biking, skiing, and rafting. In short, despite my position in society, I can determine my own coping mechanisms and the extent to which I fit into the prefigured trappings of my cultural roles.

Here, once again, we encounter the notion of individuality and freedom. Some Freudians claim that we don't have control over ourselves—over the darker sides of our bewildered psyche, subconscious, or over our identity formation. By contrast, I do believe, along with the ancient Greeks and the Enlightenment leaders, that we can and do take control over those parts of ourselves that disturb us and interfere with the way we want to lead our lives. We can cognitively control our fantasies and desires, limit our greed and anger—that is, think before we act. We can cognitively access those parts of our personality that hurt us, and we can try to understand, with or without the aid of a therapist, what it is that we really want, and what it is that we are willing to sacrifice to get it. In this sense, then, I believe that we have some control in how to present ourselves and how to define and redefine our identities as we mature and learn from our experiences. Especially with the advent of technoscience, when external pressures encounter us in quicker succession, the ways in which we resist these pressures or learn to deal with them responsibly illustrate how well we cope, how well adjusted we become.

I would like to emphasize, though, that what I mean by responsible coping mechanisms might differ from what therapists mean. For them there is the issue of deviant behavior as defined by social norms, the kind of behavior we must change and avoid, eliminate from our daily interactions. For me, on the other hand, there is less a sense of needing to belong and conform, and more of a sense of understanding the norms in order to learn to avoid some of them and embrace those that seem reasonable and supportive of a community. In this sense, then, I limit myself to the health of an individual in predominantly cognitive terms (unlike therapists and psychologists who pay much attention to our emotional baggage). My concern is therefore defined in terms of how we critically think and rationally articulate to ourselves and others who we are and how we would like to be accepted. It is a concern with ethical demeanor and behavior that retains the dignity of humanity, acknowledges the pitfalls of the human condition, minimizes our follies, and celebrates our strengths.

Perhaps one model of my approach could be the logotherapy articulated by the psychologist and therapist Viktor Frankl, who immigrated to the United States after World War II. As far as he was concerned, our search for meaning was as much an intellectual activity as an emotional struggle, a way to be moral regardless of the standards of morality of his day. Having lived as a Jew through concentration camps during World War II, Frankl knew firsthand what it took to survive, what it took not to be defeated by the surrounding conditions of humiliation and starvation. The ugly underbelly of humanity was exposed in the destruction of innocent civilian lives in concentration camps and gas chambers. The unbelievable happened, and some survived to tell what it was like. The search for meaning, according to Frankl, is about finding the words with which to express our ideas regarding the future of our existence, despite a present situation that is horrific. The compelling thing about Frankl's model is that it does not dictate what meaning we must find but leaves it to each individual to find his or her own meaning. Finding significance in our lives, something we can hold onto and retain as a way to sift through the troubles and anxieties we may experience, is a powerful tool. It is powerful because it can sustain us, and it can sustain us because we are the ones who have built this tool—it belongs to each of us alone. I am attracted to this model because of the use it makes of the cognitive capacity of humans. It considers a trust that we can listen to ourselves and find something interesting to hear.

Frankl reminds us that too often we forget to ask ourselves questions because we are scared that our answers will be ill formed. We forget that just as much as we must listen to others, we must also listen to our inner being. When we dig deep down we may find a treasure trove often not discovered. Incidentally, therapists can be (and at times are) helpful in walking us through an emotional maze in order to get to a level of personal discovery that is enlightening and that anchors our identity. But this exercise can be handled alone, with a bit of introspection and with the simple phrasing of fundamental questions, such as: What is important to you? Why is this value more significant in your mind than another? Who do you want to become, and why? What difference do you think you can make in this world? What bothers you most, and how do you intend to overcome this obstacle or nuisance? Though we might be able to come up with only incomplete answers, we still can use them to move to another level of inquiry, another level of self-discovery. Why not push ourselves to bring new meanings to our lives? At least these are our own answers, answers we can commit to and stand behind. These are answers, in short, only we can know for certain and for which only we are responsible.

From the personal level to which Frankl takes us we can move to an ethical level that affects others, society as a whole. My sense of personal responsibility leads me to appreciate how my thoughts and actions affect others, as well. Am I empowering my students or imposing my views on them and thereby undermining their self-confidence? Am I providing them with powerful tools of intellectual and personal growth, or am I insisting that they repeat my views and my ideals? Am I setting the conditions for respectful interactions among my employees, or am I reintroducing hierarchical lines of authority and abuse? With these questions in mind I develop an approach that allows me to draw some lessons as to my relationship with others. In following this line of argument, I focus on group affiliation in terms of a membership in some self-select, chosen "club." "Chosen," in my mind, means only that one chooses to what group to belong, not "chosen" in any biblical sense (of preference and superiority). The club I have in mind can be as open ended and inclusive or as closed ended and exclusive as you wish it to be—it can be a Harley-Davidson motorcycle group or a group concerned with environmental issues. There are no prefigured ways of thinking about your group affiliation, since you might be affiliated simultaneously with more than one group. You might be a father and a son, a teacher and baseball player, an academic and a real estate developer all at once.

However, my preference would be a model of affiliation much more in line with a late Viennese philosopher who eventually taught at the London School of Economics in the latter part of the twentieth century, Karl Popper. In his notion of an "open society," that means that the threshold for entry is as low as possible, that the gatekeepers are as welcoming as possible, so that barriers are lowered if not completely eliminated. Anyone should be able to be a college student (if no prerequisites are expected), anyone should be able to start a new business (and get a bank loan without the qualification standards only few can meet, namely, those who really don't need a loan). My concern with openness and mutual respect and responsibility is another way of speaking of friendship. The very notion of friendship, at least as Aristotle understood it, has to do with altruism and concern for others, with commitment and compassion, rather than with self-interest. Once the notion of responsible friendship is brought to the fore, it makes perfect sense to appreciate the multiple roles we each play in our lives.

For example, consider a single mother of two who is a lawyer and also a daughter who cares for her ailing father, while passionately involved (whenever possible) in outdoor sports. For her there are too many variables to simplify a definition of self but enough to ensure a sense of individuality and self-realization, enough to allow mutual respect with multiple individuals. In short, a person in this position inevitably sees herself as a member of multiple groups, multiple clubs so called, all of which have different demands on her time and energy. Yet, each one of these commitments provides an opportunity for her to express her sense of self and her sense of responsibility. Incidentally, when she does all of that, she has incorporated a sense of order in her life that is amazingly intricate and complex. But, when everything functions well, she has no problem explaining exactly where lines are drawn, where boundaries are set, and where her comfort zone is being threatened and by whom.

What I am thinking about is the way in which we discover and mold our identity, the way in which we clarify to ourselves the values we hold on to dearly and those we can leave behind. Our family and friends reflect who we are. They tell a story about ourselves we might not be admitting to ourselves. They might be revealing something about us to which we pay little attention. They might be helping us to realize some things about ourselves we usually don't care to discuss. Put differently, they order our lives in certain ways or provide an order of sorts for us as we make our way through life. They also hold us accountable for our actions and our thoughts, for our ideals and convictions. They demand of us to be responsible for ourselves and not blame the entire world on whatever goes wrong in our lives.

Likewise, our technoscientific culture defines most of the parameters of our daily existence, from transportation to communication. As such, our culture defines, orders, and develops our sense of powerlessness and control. There are things we can and can't do in our lives in terms of our buying power or the access we have to certain technological tools, like the Internet. Even when our culture seems to limit our access or puts too much pressure on us to excel in areas that are of no interest to us (such as computer literacy and the handling of automated bank transactions), we can resist the pressures. We can divert our attention to areas that could use our skills and inclinations more readily (nursing homes, child care, art), even if the social or financial rewards are not as great as other areas (computer engineering and telemarketing). It is within our power to transform our attitude toward cultural pressures, so as to feel better and be more liberated. Under these conditions, we might even be more appreciative that our friends are critical of us, steer us in the right direction, help us deal with issues we may conceal, or offer advice where none was requested.

Thinking about the way our culture pressures us, the way we respond to pressure, and the support network we develop along the way brings me back to the friendships and the myth of freedom. However free we'd like to be, we are still defined in terms of others, and we are still bound by circumstances over which we have little or no control. So, when we speak of freedom we might as well speak of circumscribed freedom, about the boundaries of our lives, about the limits we accept as necessary tools for flourishing. We might also add to these boundaries the people that either hinder or assist us in increasing our exercise of the freedoms that are under our control. These freedoms include, but are not limited to, the kind of decisions and choices we make about the prospects in our lives and the people with whom we affiliate. Only if we feel free to choose (even if these are bad choices) can we more readily accept the responsibility that comes with our choices.

The myth of freedom, then, is not so much a myth about having freedom at all but rather about what kind of freedom might be acquired and maintained in the twenty-first century. As far as I am concerned, the notion of freedom must move from the theoretical domain, the domain of language and ideological promises made by politicians, into the practical domain, the domain of daily technoscientific existence that we all experience. Once it is moved into the practical domain, freedom will not be a myth anymore but something we live through, fight for, and argue about. It will become, in one way or another, a way we transmit our thoughts and desires, a way to test the boundaries of our existence, and a way we test our relationships with our friends and colleagues. From yet another perspective, the acquisition of and debate over freedom can help reduce stress and personal frustration. When you realize that freedoms are acquired as much as given, that you can participate in gaining more freedom regarding your own choices in life, your expectations will change and your frustration level will diminish. In this sense, the myth of freedom might be useful for motivating people to fight, but devastating for people who live an uncritical, mundane existence—devastating, I should add, because of the false hopes that accompany such a myth and the eventual realization that it's a myth after all.

We come now a full circle in the Freudian context of understanding the relationship between individuals and their society. Civilization is made of individuals and the groups to which they belong. Some group affiliation is voluntary, some mandatory. In some cases we can escape our background or upbringing, in some cases we cannot. When therapists promise the world to their patients, when they suggest that anything can be accomplished with the aid of therapy, they promise too much. They in fact perpetuate their own myths of healing and recovery. What they and we must do instead is offer hope in the most realistic way, explaining what is and what isn't within our control. As the Stoics taught us, even those things that are outside our control, such as the weather and our height, can be approached in a controlled manner. As mentioned earlier, we can control the attitudes, reactions, and feelings we have about those things outside our control. We can be happy about the rain and about the sunshine; we can have a positive attitude toward situations and circumstances that we cannot change. When we focus our attention to our dispositions and not our positions, we gain a modicum of control, at least over ourselves.

In summary, the myth of freedom is a necessary component of our modern, democratic, and capitalist society. This component paves the way for us to realize our own roles in society and the responsibility we are willing to undertake in this society. This should not be taken for granted or assumed to be there under all circumstances for every individual. On the contrary, in order to assert our individuality and maintain our basic freedoms we have to fight just as much about the limits to our freedom as about our self-perception and attitude. My advocacy for critical intellectual self-reliance is not meant as a call for full autonomy, without regard to those people around you. Rather, my advocacy depends on friendships and relations with other people who can give us the feedback we need to know how far we can or can't push the limits of our freedom. They can also help us know if we have pushed too far or not far enough. In short, they can gauge our progress on the way to happiness.

CHAPTER FOUR

~

The Century of Ambiguity and Anxiety

Existential Angst

Though the term *angst* is German and the term *ennui* is French and therefore both are European imports into our vocabulary, they are useful symbols to describe a deep anxiety some people have about their very existence. To feel at a loss, confused, disorientated, and alienated is something quite common even to those who are fairly content in their lives. But to have such feelings permeate our every waking moment, affect every facet of our lives, is a much more overwhelming experience. When we feel a sense of angst in our lives, we know that we have been touched deeply by our fears and anxieties to a level that cannot be overcome simply by seeing a therapist or (for some) going on a shopping spree. We know that our lives are affected, that we have lost a sense of meaning, and that we can't find a way out of our predicament.

When we recoil at the Czech writer Franz Kafka's stories and novels, we recoil at his incomprehensible setup, and of course at the unfairness associated with such a sense of incomprehensibility. Why was K awakened by plainclothed police officers? What was his alleged crime? What should he do to exonerate himself? How can he return to live his normal life? Kafka's short stories and novels push the existentialist sense of dread to a level of absurdity and futility beyond human control. K never quite figures out what he was supposed to have done to deserve being charged as he was. He never quite gets an answer to his question about what charges are leveled against him. Since he has no idea what he was charged with, he never quite understands what he should do to defend himself. When things don't make sense, when our common sense fails to grasp what's going on, what we are really saying is that we cannot find any rational or reasonable order in what we see or experience. It is then, also, that our cognitive dissonance leads to an emotional turmoil. In short, to be cognitively at peace and to feel free is to subscribe to some level or semblance of order and control in our lives. It is what scientists strive for in constructing models that provide an explanation and have predictive power.

When we feel discomfort about our lives and our own expectations, when we are confused, it might be related to a certain friction or tension between what we perceive to be the order of our lives and the variables that don't conform to that order. In other words, there is something missing or awkward going on in our lives, and we would like to know what it is and what we can do about it. If you are working hard and performing everything expected of you as an employee or a student, why aren't you rewarded (with raises, promotion, or grades) as expected? What I'm trying to describe here is similar to the uneasy and bewildering feeling K (Kafka's character in *The Trial*) has when he is confronted, out of the blue, with a mysterious accusation about a crime he seems to have never committed. Kafka's characters are always in this daze of confusion when the normal order of the world has fallen apart in front of their eyes; they are in no position to rebuild it by themselves.

In this chapter I attempt to illustrate in more pedestrian ways what Kafka in so masterly a way illustrates in his literature, namely, that our personal feelings and mental or cognitive apprehensions have something to do with a much bigger picture of the world that influenced the development of the twentieth century. In this sense, then, our confusion is justified. Not only isn't it our fault that we feel the futility of our position in life, it would make no sense for us to feel otherwise. This framework is technoscience, the combination of the theoretical and practical forces of science and technology. It is the way science and technology have been treated and have treated us that actually influence every aspect of our daily lives and therefore also our psychological dispositions.

To contextualize the existentialist concern with fear, dread, and futility in the past century, it would be useful to briefly recount the technoscientific development of the twentieth century, from the Logical Positivists to the Manhattan Project. The sketchy historical survey focuses more on some scientific and technological highlights (theoretical and practical alike) than on a full survey of the historical data, because of the assumptions we associate with science and technology. These assumptions include items such as the regularity of natural phenomena, the predictability and certainty of daily occurrences, and the ways in which we can rationally explain almost anything we observe or experience. The citadels of certainty have been questioned in the past century, so it is not surprising that our culture feels less secure about its future than it did one hundred years ago. One can't but observe that it's necessary to address certain cultural facts and artifacts regarding technoscientific development with human perception, understanding, and reaction. The occurrences of the twentieth century, such as world wars and mass destruction, turned our Western culture into a culture of ambiguity, anxiety, and anguish, and influenced the lives of most thinking people. How could you avoid worrying about the extinction of the world? How could you ignore the uncertainties and predictions of doom that permeated the popular media? How could you not be personally affected by these big events?

Perhaps I should distinguish a bit more carefully between basic human fears of survival and what we consider our doubt concerning science and technology. We can be critical and concerned, doubting the details of every piece of information that we come across in the media. This kind of doubt, the ongoing questioning of everything and everyone, the doubt we have about the authorities in politics, medicine, and science, is part of the progress of our culture (and how we become wiser in appreciating it). As we become more educated through institutions of higher learning and mass communication (the press, print media, television, and movies), we expect those in power to answer our questions concerning their judgments. When we doubt them, it is not because we mistrust them or ourselves. Rather, it is an indication that we are concerned and that we are thinking critically about the information transmitted to us, whether it's about water shortages, an impending energy crisis, or the dangers of pollution.

So, doubt, in my mind, is a constructive element in the exchange of ideas between an elite group of experts and the general public, a way to communicate in lay terms what we should know and what we can do about a situation. Fear is another matter altogether. Though routinely associated with doubt, fear is less intellectual than emotional. Our fear cannot always be explained away, since it might be completely irrational. Fear sometimes gives us an intuitive insight (a gut feeling that something is wrong) we might miss otherwise, but at other times it can be rationally explained and understood. Fear is a powerful motive for reconsidering what we do and how we go about doing things. Doubt and fear are part of the quest for an open-ended order and a way for us to act responsibly in the world. It's because we are fearful that we would like to know more, rather than less about our surroundings. We want to understand why we are fearful and what we are fearful of, and what we can do about reducing, if not completely eliminating, that fear. If a good explanation is provided, it could help reduce our fear. Perhaps we could even be better prepared to conquer our fear; perhaps we could even predict under what conditions it arises and find ways to circumvent it.

Our fears and anxieties lead us to seek explanations. With appropriate explanations we get a sense that we can organize and control our lives, and leave behind our confusion and fears. For example, if we know that our own energy conservation can make a difference to the entire state and that it can contribute to the reduction of rates, then we are more likely to conserve. It's because we have anxieties that we yearn for definitive answers (even if they are numerous and require making choices among them), ways to order our world and our decisions in life. Now, if we could appreciate where the fear and doubt are coming from, if we could trace their genealogy into our minds, then we could reduce if not completely dissolve them. This is where my retelling of the story of technoscience in the twentieth century comes into play. In looking at its development, we see that what seems a personal, individual anxiety is in fact connected to a culture of anxiety. Our feelings, in this context, are not irrational at all, according to my view, but are based and substantiated by the material conditions that surround us.

One more preliminary remark: In the previous chapter I made the connection between the expectation of freedom we share in the technoscientific Western world and our quest for some semblance of order and control in light of this situation. If freedom is linked to our sense of control, and if our sense of control depends on the things we can explain, predict, and interact with, then it makes sense to speak of cognitive order—the order we establish in the mind. The order I focus on here is personal and intimately connected to our immediate surroundings. This means that we want to have a steady job with specific duties and responsibilities; that we can expect to receive specific rewards, financial and otherwise, when we perform well what we are supposed to do; and that we expect to get raises as we progress up some corporate ladder. Likewise, we expect our family relations to be organized around some principles and values we understand; we expect our friendships to be reliable and responsive to our needs and overtures. Finally, in the more general sense of our daily existence, we expect a legal system that is fair, explicit, and open to public scrutiny.

Twentieth-century existentialists worried about the incessant shift over the previous century toward a bureaucratic existence that confines our freedom and dictates our every move, one that loses sight of the individual. When the twentieth-century German sociologist Max Weber described the onslaught of institutions and their instrumental rationality, the way they articulate and justify their own structures and the rules by which they insist we all abide, he lamented the loss of personal relationships and the empathy that is associated with human interaction (1968). The shift to a highly bureaucratized culture has its advantages, of course. We call those productivity and efficiency, fairness and equality (of treatment). Incidentally, these advantages were envisioned by the Enlightenment ideals of equality and liberty. We have elevated these concepts into equal opportunity to all (at the end of the twentieth century), without realizing that there are costs to these advantages. The price we pay for equal treatment is the price of anonymity. In order to be equal before the law, justice must be blindfolded, not see who it is that comes before the bench. In the name of fairness we have depersonalized our culture and lost the compassion and pity that Rousseau claimed to have perceived in his Noble Savage (his idealized view of humans in the state of nature, before education contaminated their care about other human beings). Unfortunately, this has meant equally poor treatment of all, complete disregard for individuality and unique differences. We get lost in the system, just like Kafka's characters in the bureaucratic labyrinth, with stunned looks and a deep sense of futility. Just think of the times when you go to get your driver's license, when you register for courses, or when you apply for any government agency's program.

In the course of the development of technoscience, we can observe a shift from ambiguity to anxiety and from anxiety to anguish. The technoscientific community appreciated and acknowledged, over time, the inherent and inevitable ambiguity of its trade—anything we thought was true might turn out to be false. Whatever model we constructed of the world is found out to be wrong and in need of revision. For example, we had a geocentric view of the universe, where the earth is at the center of the universe, but it was replaced by a heliocentric view of the universe, where the sun is at the center of the universe. By the twentieth century we had adapted a relativistic view of the universe, thanks to Einstein, where planetary constellations are in relative positions to each other, and where it makes little sense to speak of a center of the universe any longer. Newton's view of some absolute or real mass (weight) has been challenged by Einstein's theory of relativity and by quantum mechanics. Whom should we believe? Who is right? Is the latest indeed the best (most accurate) representation of the universe?

Ambiguity leads to a mode of thinking that is in constant search for additional alternatives and answers to given problems. It is this sense of critical self-evaluation that has marked both the acceptance and the success of technoscience all along. Ambiguity, then, forms the basis of the development of the technoscientific discourse (which includes both theoretical discussions and laboratory work). This is the case first because of the introduction of radical doubt (by Descartes in the seventeenth century) about what we know. Secondly, this is also the case because the method of radical doubt insists on an ongoing search for better, more accurate answers than those we have had acquired so far. The ambiguities we seek out and learn to live with turn into a more comprehensive worldview that suggests how inherent they are to our process of knowing. But if this is what we have to live with, would we ever find ultimate answers to our questions? Will we ever know for sure how the world works? Are we bound to remain anxious about the trustworthiness of what we know? Are incomplete descriptions of the workings of nature indeed explanations?

Anxiety, then, is the necessary (if not sufficient) fiber with which "progress" (or even the vitality and survival of the technoscientific community) is guaranteed. Without this ongoing tension of the inconclusive nature of the technoscientific project, this project would come to a standstill, with no one interested in pursuing yet another unresolved mystery or anomaly. Finally, there is a deep anguish to be found in technoscience, an anguish that some try to either cover up or ignore, and that some, like Gerald Holton, characterize as "despair" (1996, 156). If, for example, more attention were paid to the anguish of technoscientists during and after World War II, it would be clear that anguish is not the exclusive domain of art and literature, as some would suggest. This does not mean that the horrors of World War II would have been averted, but it may have forced military and political leaders to reconsider their wartime proposals.

Technoscientific Ambiguity

When something is ambiguous we believe it either to be obscure or to have more than one meaning. This is true in our daily lives and in nature as a whole. There is a certain discomfort when we feel surrounded by ambiguity; we like clarity and order, to know the rules of the game we are playing. Since life can be somewhat overwhelming and frightening, any level of reassurance would help ease our daily concerns. There is a relevant story about what happened at the British Association for the Advancement of Science when it appointed a committee on technical and scientific evidence in courts of law in 1862. The committee reported that in many legal disputes, the testimony of experts was contradictory and confused juries and judges alike. In doing so, these experts undermined the credibility of their professional testimonies. In order to support the reputation of scientific and technical experts, the committee suggested that judges be empowered to appoint expert assessors who would review expert testimony and offer their own assessments. Who would ultimately decide on the credibility of expert testimony, the experts, the assessors, or the judges? What standards should be applied? (For more on this see Sassower 1993.)

When in doubt, some of the elders suggested, appealing to someone outside their circles made sense. But how would their judgment be any better? Where does "the buck stop"? These questions came about not only because of the contradictory nature of the testimonies but also because the legal system expected it could rely on clear and unambiguous testimony of scientific experts. We might accept contradictory legal interpretations of the facts, but can the same be said of scientific interpretations? Is there no solid foundation on which we can all agree? These questions are quite old by now. Yet, they illustrate the recurrent problem of resorting to the advice of experts. Experts are supposed to know for sure, to have no doubts, and to be able to figure out exactly what is going on and what needs to be done. This was supposed to be true especially in the case of science, but even in this "exact" field of inquiry this wasn't the case. The British courts, 150 years ago, recognized the difficulty of appealing to experts. Has anything changed? To some extent, nothing has changed. Judges still revert to the advice of experts in product liability cases as well as in medicine and hope to achieve a level of legal certainty when rendering judgment (a level, incidentally, that was given up by the technoscientific community a long time ago). Perhaps what has changed is the legal climate in contemporary culture when we believe that we can "buy" expert testimony and manipulate the facts to suit our purpose. Is the age of ambiguity turned into an age of cynicism? Can we count on anyone anymore? Is expert testimony tainted, or is it simply too inconclusive to be counted on?

As the British tale shows and our perceptions of the world reaffirm, ambiguity arises not only because we don't know what to make of this world but more so when we receive more than one answer to the question we ask about this world—for example, competing answers about the safest and most efficient energy source (coal, nuclear power, solar energy, oil, natural gas, wind). The great influence and appeal of church doctrine has always been its straightforward view of the world. There are good and bad things, there are good and bad people, there are things for which one goes to heaven and those for which one goes to hell. Boundary lines are clear-cut, and when you transgress, there is a local priest to judge you one way or the other and to determine your punishment. Most religious institutions understand that their major task is to alleviate the fears of their members. The method used by religious institutions has always been underlined in a doctrine or a sacred text, like the Bible or the Koran, that provides a model of an ordered universe with divine purpose. In a well ordered church hierarchy, every member knows her or his position and for what they are responsible. In short, the quest for order could be satisfied through religion. But what happens when the authority of the church loses ground in the face of the emergence of science?

To some extent, nothing has changed. The image of the all-knowing priest with his long robes and esoteric language (Greek, Hebrew, or Latin) transformed into an image of scientists and engineers in white robes, using mathematics and computer code as their esoteric languages. The one group remains as aloof and incomprehensible to the general public as the other. The authority of the one is just as daunting as the other, and the sense of awe promoted by the one is not diminished with the appearance of the other. We teach theology as well as science, but the insights of the chosen few remain beyond daily comprehension, whether in one field or the other. We appeal to scientists just as we appealed in the past to religious leaders—we want them to explain to us why something happened (a hurricane, the death of an infant) and predict what will happen to us in the future (will I be rewarded with a healthy, long life for good hygienic and nutritional behavior?). The mysteries surrounding God have been replaced with the mysteries of scientific knowledge (subatomic particles, genes, even the movement of the stock market). As much as we feared God's wrath, we fear nowadays the wrath of science, so we turn our faith to those we think can be the intermediaries between us and the abyss of the unknown. We want their deliverance from our anxieties to heaven on earth: secure knowledge, predictable chains of cause and effect, and ultimately personal tranquillity. Can the new priests deliver the promise of happiness where the others have failed? Can the new priests be as successful as the old ones in promoting their worldviews?

When the technoscientific community began to establish itself, with the success of the scientific revolutions of the seventeenth century (and many subsequent ones), it could be seen as imitating religious institutions (even though there was rivalry between them). Scientists prided themselves on being able to organize, explain, and predict everything under the sun. Theories and models were so comprehensive and absolute that there was no room for ambiguity: this is how the world works, this is what gravity does to us, this is how gasses behave under pressure, and so on. Modernism and the Enlightenment of the eighteenth century have retained their appeal because of their promise to reduce ambiguity and bring to each one of us an ordered view of the world (such as the Cartesian geometrical coordinates and Newtonian mechanics). Doubt can be replaced with certainty, if you are willing to follow a few mental and physical steps, they said; all you need is an open mind, a keen eye, and a rigorous method of analysis.

For example, the seventeenth-century British natural philosopher and politician Francis Bacon showed the way to fight the "idols of superstition," as he called them, and turn humanity from belief in unsubstantiated claims to a belief based on observations. He added the requirement that we experiment and test every claim that comes before us so that we know for sure if it's true or not. He listed four sets of "idols" that obscure our perception and interfere with our knowledge. The first are the Idols of the Tribe, which he identified as our human follies when measuring our findings in terms of our own senses. The second are the Idols of the Cave, which are our personal follies that distort our perceptions, the subjective perspectives we employ when trying to make objective judgments. The third are the Idols of the Market, which are identified with our cultural biases and the obscurity of the language we use to describe natural phenomena. The fourth are the Idols of the Theater, identified in terms of the systems we have constructed through our tradition. Like the previous set of idols, these too contribute to the misrepresentation of our perceptions of nature. Bacon's warnings set the stage for a more objective, value-neutral observations that can be tested empirically and then generalized (1985, 48–49). Bacon, in fact, shifted the level of debate from the open-ended superstitious beliefs of the old ages to the tried-and-tested experiments of a new age. Every claim had to be demonstrated publicly and could be tested again and again by anyone who wasn't sure whether to believe the claim under consideration. One didn't need to trust the word of some person in a position of authority but could find out for herself or himself if a claim was true or false.

Open and read the book of nature, Rousseau told us (concurring with Bacon's recommendations), and not some obscure religious book. Unlike a religious book, the book of nature is available to anyone who wishes to read it; all it requires is a mind open to accept that which is readily there. In fact, as Bacon, Descartes, Kant, and Rousseau reminded us, there is nothing mysterious about nature (and therefore nothing mysterious about science). With careful methods of inquiry, using both our reason and our senses, we can discover anything we are interested in. We can know for sure how the world works; we can stop speculation and superstition; we can test any claim and confirm or disprove it for ourselves—we need no intermediaries. The belief in human common sense and intelligence to explore the mysteries of the world, and the conviction that we are all equal to the task at hand, were fundamental principles that made science so popular in the past (and that still inspire young and curious minds to become scientists). The radical switch that took place some three or four hundred years ago is taken for granted today, but it was radical nonetheless. It was radical because it allowed every individual to participate in the scientific enterprise and contribute to its success. It tried originally to demystify our knowledge acquisition, even though in later centuries the mysteries came back into play as the reasons for the authority of experts. Once intermediaries are needed, there are barriers to direct access to knowledge. These barriers are set up as power relations among people and eventually create a secular hierarchy that parallels a religious one. But before we go on to this subject, let us remain with the great promises of the modern era in science and technology.

The Enlightenment promise has been fulfilled, according to scientists, because the order of the universe as presented by them is based on direct observations of nature and a mathematical system of logical deduction rather than on belief and faith, superstition and miracles. Though this order has changed over time, there were good reasons for the changes, all of which can be explained and documented (from gravity to blood circulation and the human immune system). New observations of celestial movement, in Galileo's time, or new calculations of the relative position of planets, in Einstein's time, compelled us to rethink what people once thought was right. We've changed our views not because someone told us to do so but because the evidence they provided convinced us that they were right. This evidence wasn't some esoteric mystery; it was available to people to weigh on their own. Moreover, unlike religious communities, the technoscientific community boasted that anyone could challenge anyone, that anyone could propose alternative explanation and models. For instance, a British outsider called Faraday made a name for himself with his discovery of electromagnetism, a Jewish clerk in the patent office in Switzerland (Einstein) challenged the big boys in fancy universities and laboratories, and a quadriplegic in England (Stephen Hawking) proposed new models for the conception of time and space. So, the authority of anything we say about the world remains human and therefore open to refutation (as Karl Popper suggested in 1959). Our models of the universe become targets for scrutiny, testing, and falsification. If you have something better to offer, the world will listen to you. Any appeal to divine revelation or the power of tradition is viewed with skepticism, if not outright scorn. Any attempt to bully scientists, even if it succeeds for a while, blows away when fresh evidence is contrary to the established view. The truth is more powerful than any institution, and it will undermine any claim to authority by anyone. Can the same be said about the authority of the church?

One of the problems with this scenario is that science didn't do a great job of explaining itself to the public or at replacing religion, which many Enlightenment thinkers sought to do. As I said earlier, the church provided a complete picture of the world that made sense, that gave direct instructions about what people should and should not do. You could argue with this or that aspect of the doctrine; you could even argue with your local priest. But by the end of the day, the priest still came up with yet another explanation and yet another principle that you should have known already, another reason why you had to accept everything that was presented in the name of the church. Why did you have to accept what the priest said? Partly because if you rejected one part of the religious doctrine, the whole doctrine fell apart, and that was too much of a loss for anyone to bear. So, you let some ambiguities and questions persist and buried them deep down in the recesses of your mind. But isn't this the same when we speak of the scientific model? Not quite.

The philosopher of science Thomas Kuhn suggested that one paradigm of the scientific model, one picture of the world, is replaced with another, as more discoveries come to light (1970). He called that the revolutionary character of science, that which propels scientists who are at a dead end to come up with alternative frameworks for the data they have collected. Kuhn's view helps explain how and why scientific models change radically, and also the unpredictable path scientific developments have taken over the centuries. Yet, one of the unintended consequences of his view is that instead of believing that the path to scientific progress is clear and cumulative, we might suspect that it is arbitrary, even capricious. Can we predict the next paradigm shift, the next time a whole scientific model is completely replaced? Could we have anticipated the shift from Newtonian mechanics (where mass, for example, is understood as a constant and absolute factor) to quantum mechanics (where mass is understood to be relative to its velocity)? Probably not. We needed first to face some puzzles that we couldn't solve the old-fashioned way, that we couldn't solve at all.

Another factor that figures into contemporary science and how it's played out in our culture is the relationship between the scientific community and the economic and political environment in which it operates. One example that comes to mind is the famous story of the Supercollider Superconductor, which was designed in the 1980s but was never built (see my book on this subject of 1995). While there was an argument to be made about the scientific value of building this monumental project in the Texas wasteland, some even compared it to the Egyptian pyramids; there were too many financial and political obstacles in its way. What ruled the day, when everything was said and done, with congressional hearings and the testimony of experts, was the sheer cost of close to ten billion dollars and the lack of political support. A scientific enterprise, like any other in our culture, isn't isolated or protected. It requires the kind of political savvy and financial commitment that athletic teams need when they try to relocate to another city. Perhaps this is so because science, unlike religion, cannot count on the faithful for donations. It must appeal to the corporate or national security world to fund itself. The lesson to be learned from this experience is also that our faith in technoscience is weaker than our faith in God. We have heard too many stories that didn't pan out; we have seen too many accidents (in the case of nuclear power plants) to be more trusting of the proclamations of scientists. Also, for some odd reason we still distinguish between technoscience and religion. The appeal to an omniscient God is more attractive than one to an all-knowing expert who speaks in esoteric terms about bizarre ideas.

Surveying just a few of these stories of the past century has made it clear to me that our scientific models have added confusion and ambiguity rather than provided a final statement on how the world works. You see, it seems to most secular observers that religious leaders never waiver in their belief and explanations; they invoke divine intervention or detachment when and however applicable. When we consider popular media presentations of religious leaders and their practices it seems that they appeal to God's mysterious ways when faced with incomprehensible situations (the death of innocent children). But somehow they manage to promise more than they can deliver and still retain their appeal. They counter the existentialist view of the absurdity of human life with an insistence on divine meaning that is only partially spelled out to the masses. When in doubt, offer an answer. When the answer is questioned, offer a leap of faith. When the leap is refused, offer damnation—scare the doubtful into acceptance. If the skeptic still resists, repeat the process or excommunicate. The question remains: can technoscience use this method of persuasion?

We are more sensitive today about teaching the history and sociology of science rather than just the success stories of one invention on the heels of another. When the history and sociology of science are taught, they tell a very human story of success and failure, of competition and greed, of how a community of people works together and tries to produce a model acceptable to all. Whether we speak of the theory of relativity or quantum mechanics, whether we ask about the intricate structure of atomic particles or the complexities of galaxies, we are multiplying ambiguities rather than reducing them. We have to interpret and weigh alternative answers. When we interrogated American or German scientists during and after World War II and asked them about the development of the atomic bomb, we heard conflicting stories. We hear of how a group of scientists were put together in Los Alamos under military supervision to defeat fascism. How did they feel about their work? How did they reflect after the war on their accomplishments? Did they achieve the goals intended for them? Did they do more harm than good in killing Japanese civilians? The answers they give us are as perplexing as our questions. The conflicts to which they admit put a strain on our comprehension. The views we form along the way don't provide a clear-cut picture of what happened or what should have happened. (For more on this topic see my book of 1997.)

If we looked at science as the ultimate modern replacement of the dogmas of religion (those ideas and rules that can never be changed because they are God's words), then we will have found two related things we didn't quite bargain for. Perhaps this is so because we are dealing with people and their personal anxieties, hopes, and dreams. First, science and the scientific community have the same tendencies that religious institutions have—they are likely to become authoritarian, narrow-minded, and orthodox. They tend to initiate their members into a framework that must be accepted and to which they must adhere (see Thomas Kuhn's description of what he calls "normal science," the way the scientific community operates when there are no major problems or obstacles). When they are trying their best, they allow some deviance and challenge, some critical distance so that new venues can be explored. Second, because the fabric of the scientific community allows some deviance, there are many cases where the theory of yesterday is proven to be wrong and is replaced with a new theory (the earth is the center of the universe; no, the sun is the center of the universe; no, there *is* no center of the universe!). When this process of change happens, when the explanation we give and the models we propose are constantly in flux, there seems to be no foundation at all. Technoscience seems incapable of giving us a final verdict on anything! The ambiguities of yesterday are replaced with the certainties of today only to become themselves ambiguous enough to be replaced once again.

There has been a shift from the scientific revolutions of the seventeenth century and even the promises of the Enlightenment leaders in regard to the attainability of certainty. Nowadays we speak of probability and putative knowledge claims, as opposed to certainty. Perhaps the great conceptual changes and the discovery that yesterday's truth has become today's falsehood sobered technoscience into making weaker claims about its ability to provide foolproof models and theories that explain nature and predict future developments. The scientific community portrayed itself by the end of the twentieth century as if no certainty could ever be achieved, and even if achieved, it would be short-lived. This is in order to stress the dynamic nature of the scientific inquiry and the open-mindedness of all participants. The "book of nature," as it has been called, is open to numerous interpretations and revisions;

it remains open for every new generation to seek its knowledge and understanding from it. (An interesting parallel can be drawn between the book of nature and the Bible in regard to openness of interpretation in relation to orthodox views that disallow multiple interpretations, and those that encourage them.) We learn to read the world differently all the time, and this is both exhilarating and frightening. The exhilaration comes from a sense of adventure and satiated curiosity as we invite ourselves to yet another feast of facts and experiences. The frightening part of this process relates to our sense that we'll never quite get the whole story and that whatever we believe to be the case today will turn out to be false tomorrow.

This reminds us of the inherent ambiguities we must live with, the uncertainty regarding what we know about the world and humanity. Under these circumstances, we are not escaping the fate of our ancestors who worried about how little they knew of their surroundings and wondered how they would ever be able to face natural disasters. Therefore, it seems to me that the responsible quest for order remains strong in all of us, whether we admit it or not. Of course, the quest for order itself has changed over the years under the influence of the technoscientific community, because the concept of order itself has been challenged. For example, we may talk of an open-ended order, where alternative ways exist to establish and construct a model of order. Having options from which to choose may seem contradictory for those who seek a simple and straightforward order, but having these options remains a way of allowing us to make choices, to control the variables with which we order the world in more than one way.

From Ambiguity to Anxiety

We learn to tolerate ambiguities, because we know that there are different ways of looking at the world. When we say that a certain food is nutritious, we accept that forthcoming reports may dispute this statement. Competitors might dispute the claim, government agencies might study the claim, and our neighbors might tell us of their positive or negative experiences. In short, we know not to trust anything that is promoted by corporate America, because the commercials often promise more than they deliver. Advertisers have a stake in increased sales and profitability, so consumers' well-being might be secondary on their minds. When we add science and technology to this situation in the hopes of reducing ambiguities, we might find ambiguities rising when we expect them to diminish.

The technoscientific enterprise is committed to minimizing ambiguities so as to provide a better approximation of the truth about the world as we know it. The goal, as Popper and the Vienna Circle understood it, is to positively know enough things about nature to negotiate our survival. Originally they thought that the more data are collected, the better the generalization from that data, and the closer to the truth we can come. But is the generalization foolproof? Can we take the next incident or occurrence and know for sure that it will turn up just like all the previous ones? Eventually, these philosophers of science realized that perhaps there will be incidents when the general rule doesn't apply. So, they changed the formula to say that when we know for sure that a hypothesis is false, we know for sure that this is not the case, and then we know quite a bit. We may postulate supporting ideas (auxiliary hypotheses), revise our original hypothesis, and eventually test a better, more critically articulated hypothesis. The expectation is that in the process we will learn how to avoid errors (or generalizations that don't fit all the data) and learn as well to venture into territories that might provide further hints about our reality. The "logical positivists," as they were called, tried to get as close as possible to the direct expression and documentation of reality, collecting, analyzing, and building piece by piece as much data as possible. Their belief was also that the appropriate use of language would separate those claims that are metaphysical (untestable, vague) from those that refer directly to reality (verifiable, confirmed, falsifiable). Some of them tried (Wittgenstein 1981) to provide a blueprint of what could and couldn't be said about the world, what one could be sure of and what one should really avoid talking about.

Whether the focus of the logical positivists was on the tools and methods of inquiry or the linguistic models that attempted to mirror nature, their enterprise was devoted to distinguishing between superstitious beliefs (pseudoscience) and scientific ones (models and systematic frameworks). Even when we only approximated the Truth, the scientific process could lead us from ambiguity to anxiety. Every principle, model, theory, or hypothesis could face rigorous testing that might show it false. Even when one sits, as it were, on the shoulders of technoscientific giants, success is not guaranteed. To some extent, this condition (of anxiety about never reaching the Truth, the whole Truth, and nothing but the Truth) is part of the technoscientific gestalt (mind-set, frame of mind, attitude, worldview). As I continue to argue in this book, this way of thinking finds its way into the popular culture not only as an acknowledgment of the inherent uncertainties of our knowledge claims but also as the inevitable anxiety that permeates and motivates the human quest for order. Kafka's characters would find themselves by the end of the twentieth century not in a legal and bureaucratic labyrinth but in a technoscientific maze where the pretense of certainty has been given up.

What makes the technoscientific gestalt unique is that it is so by design and not by default. There are good reasons to give up claiming that we know anything for sure. We can make educated guesses, provide hypotheses that can be confirmed or refuted. Claiming to know more than that is simply wrong. Furthermore, those joining this group activity are conscious of this predicament of having to assess and judge in a manner that would provide clear answers to complex questions while knowing full well that these answers can be neither clear-cut nor definitive. The most they can do for us is offer provisional answers, the credibility of which is open to further questioning. Anxiety inevitably creeps into the minds of these practitioners: What if the puzzle cannot be solved quickly, or not at all? What if the puzzle is solved, but there is no recognition of the solution? What if there is a situation where multiple solutions are offered simultaneously? What if someone steals the solution? What if funding agencies are strapped for funds and no grants are awarded in a particular field of research?

These are haunting questions for anyone being initiated into the technoscientific community. After training for years and then having to keep up with the latest global developments in the field, it's not surprising that anxiety runs high among technoscientists. I would like to add here as an aside that anxiety is associated in popular culture with the "creative" bunch among us, poets, artists, and musicians, and is seldom understood to be part of the technoscientific gestalt. All one needs to do to appreciate the error of this conception is to read some of the writings of some of the Manhattan Project participants, like Oppenheimer (1955). In their diaries and memoirs, they confess to having doubts about their technical work and about the moral implications of implementing anything they helped to produce. Are they artists or philosophically minded scientists? Are they engineers who merely execute orders or intellectual giants who carefully examine their options before making a choice? Or are they cogs in a cultural wheel whose destiny escapes their vision?

One way Stephan and Levin document the creeping anxiety of technoscientists, and even explain certain demographic realities of their community, is to focus on age. In general, they conclude that younger scientists are more productive than older ones and that "there is a strong relationship between age and the ability to do path-breaking work" (1992, 156). In contemporary society, the focus on age, and especially on the premium given to youth, produces a progressively more pronounced sense of anxiety within the technoscientific community. According to this study, every scientist is bound to do less work and less important work as the years go by. Instead of accumulating knowledge and expertise, they say, one loses touch with the most exciting and demanding developments in one's own field of research so as to remain within the fold of the technoscientific establishment and make as few waves as possible. As they fall behind, these practitioners lose faith in technoscience and the meaning of their roles in sustaining it. They are definitely different from the religious practitioners whose faith and conviction grow stronger over time.

As Stephan and Levin argue, there is another important factor that dominates the technoscientific community and that also contributes to a sense of anxiety. They call it "RPRT: being at the right place at the right time" (158). If we recall the stoics and their demarcation between those things that are within human control and those that are not, RPRT obviously falls into the category of those things over which humans have no control. If one cannot control the circumstances that will dictate success or failure, then it is obvious that hard work and dedication alone will not assure an anxiety-free working environment. How could one anticipate the next stage of digital development? How could one ever hope to hit a moving target?

Anxious practitioners are under tremendous and unproductive pressure, some of which is not of their own doing. Instead of motivating them, providing more incentives to work harder, the competitive pressure described by Stephan and Levin is counterproductive: it mystifies and terrorizes; it might even paralyze! Stephan and Levin make clear in their study of the workings of the technoscientific community that these professionals exhibit the behavior of other professionals in contemporary culture. The citadel of science and technology is no guarantee of security and stability, of peace of mind. Ambivalence and ambiguity about what we know lead to anxiety about what we can and should do about what we know. As Freud says, "If civilization is a necessary course of development from the family to humanity as a whole, then—as a result of the inborn conflict arising from ambivalence, of the eternal struggle between the trends of love and death—there is inextricably bound up with it an increase of the sense of guilt, which will perhaps reach heights that the individual finds hard to tolerate" (96).

Guilt is another term one can use to describe the kind of anxiety that torments humans. This does not mean that the two terms are identical or that one adequately describes the other; all it means is that the one is analogous to the other. This guilt or anxiety leads Freud to speak of civilization as being "neurotic" in some general sense (110). Freud brings together the concepts of guilt and anxiety and relates them to the twentieth-century development of civilization: "Men have gained control over the forces of nature to such an extent that with their help they would have no difficulty in exterminating one another to the last man. They know this, and hence comes a large part of their current unrest, their unhappiness and their mood of anxiety" (112).

The conquest of nature, as imagined for centuries by the leaders of scientific revolutions, turns into the nightmare of the annihilation of humanity. Two world wars defined and illustrated the kind of annihilation we could expect, from the war in Europe to the use of atomic bombs in Japan, from the systematic execution of civilians by the Nazis to the mass destruction of cities by conventional means. It is painful, ugly, and devastating. It crosses national borders and social classes. It disregards personal guilt and innocence, and it focuses on destruction as such. Is there a way to avoid this technoscientific progress? Is there a way to shift the march of rationality into a march of compassion? How can we responsibly deal with the gifts of technoscience?

According to Bauman, human anxiety might be mediated, if not eliminated. The "expert" is singled out as a "translator" who can simultaneously deal with the subjective anxieties of individuals and the objective claims made by the technoscientific community. The expert shuttles between these two spheres of conduct, delivering messages from one group to another, while maintaining the goodwill of both. If done correctly, this work can be useful and helpful, mitigating unintended negative consequences so as to improve the conditions of humanity and prospects of progress. Eventually, the work of experts seems necessary so that widespread breakdowns will not devastate civilization (1991, 199–230).

When experts are spokespeople or media consultants, as I have argued elsewhere (Sassower 1993), they might be suspect when they translate from one community to another. They might conceal all potential problems to make themselves understood or to ensure the transition and acceptance of the community at large (of whatever the technoscientific community thinks is appropriate). They might try to calm the public's concerns because insufficient information doesn't allow a definitive answer. Finally, they might be expected to say more and commit to more than they would like. But public pressure for reassurance pushes them into that unenviable position. In its quest for order and stability, the public expects technoscientific certainty, as opposed to religious comfort. As such, it expects explanations based on reason rather than faith. So, anything short of conclusive assertions is deemed unsatisfactory and might raise anxiety. If technoscientists don't know the answer for sure, who does? God? Are we ready to resort to the old-fashioned answers?

One could come up with additional complications concerning the feedback loop established between the public, the technoscientific community, and the expert. For example, what if the technoscientific community internalizes the fears and anxieties of the public and thereby caters to these fears and anxieties? What would be the nature of the technoscientific enterprise if it were constantly to pander to the public and its expectations? Would we not have a more conservative, less imaginative and risk-taking enterprise? Would we not have a less critical and honest enterprise, because any mistake, error, or failure would be covered up? Instead of technoscientific debates open to the public (about the Human Genome Project or the use of genetic engineering in agriculture), so as to inform the public of ambiguities, there would be debates about what to disclose and what to hide from public scrutiny (preferably before a disease is diagnosed). In the face of incessant anxiety, a condition that seems beyond reproach or repair, there creeps slowly an admission that nothing can be done to change the situation, and deep pain begins to manifest itself. This brings us to anguish.

From Anxiety to Anguish

A certain amount of sadness accompanies the appreciation of our position in life. We remain puzzled and awed by the inexplicability of the universe and human relationships. Our adventures into space, the Human Genome Project, and cellular technology—these and other technoscientific advances have increased our sense of confusion and the recognition of the limits of our own world. We proceed *as if* we know something true and believable, *as if* we have a handle on reality, and this method indeed becomes our myth of reality, our way of avoiding responsibility. I call it a myth, just like the myth of freedom, not because there is no such thing as reality (there is also something very real and realizable in freedom) but because of the claims made in relation to reality. At best we come in touch with the fringes of reality, with its surfaces and edges. We proceed *despite* the limitations of our knowledge of what surrounds us and what feels real to us, hoping to do the best we can with our lives. We navigate without a compass, hoping that the stars above and the lights along the beaches will guide us to safety.

Our predicament is a cultural phenomenon. It may be more fully articulated by philosophers and poets, by reflective technoscientists and spiritual people. But it remains a predicament we can all feel in our bones—it is this awful sense that we don't quite "get it," that we don't know how to act responsibly in its presence. The "it" we are trying to grasp eludes our comprehension, being just one step ahead of us, no matter how educated we are. There are supposed existentialists who have been openly concerned with these issues for quite some time. They have been known to ascribe futility to the human condition and are sometimes labeled "nihilists." They perceive life to be futile and absurd, so ridiculously meaningless that it might not be worth living. For example, when Camus describes Sisyphus as an absurd hero, he explains that Sisyphus is conscious of his predicament. What makes his situation seem absurd is the very possibility of accomplishing an impossible task, rolling a rock all the way to the top! But poor Sisyphus cannot accomplish his task, cannot achieve his goal. He remains in our collective minds a symbol of what we are all condemned to suffer—the humiliation of a never-ending quest to accomplish an impossible task, achieve an impossible goal. This is what we call existential *angst* or *ennui*. In plain English it means looking into a mirror and seeing nothing but a dark abyss and realizing that this where we are, this is what life is all about.

Technoscientists are no different from philosophers and poets in having a conscience, and perhaps they are in a fortunate position to exemplify the concerns of the entire culture in which they live. They are painfully aware of the human condition and frame it in a manner that induces us to realize our personal involvement and potential to overcome our condition, perhaps not in any global, universal, absolute fashion but rather in a personal, subjective, and relative way. It is the contention of this chapter that technoscience does not and cannot fulfill our quest for order. Since technoscientific theories and practices admit to the inherent ambiguities they contain, they end up inadvertently raising personal anxiety. The anxiety is not necessarily a feverish one that turns into psychotic behavior, but it could be pushing us toward a cultural anguish of sorts. If there is a quest for order, it is an open-ended quest, one that does not prefigure what kind of order is needed or would be preferable to any previously stated order. The order we seek might be informed by technoscience but require a perspective and an application broader than technoscience. As I mentioned earlier, though scientists appeared to replace the cultural position and authority of priests, they ended up losing the confidence of their audiences because they admitted too many ambiguities, too many anxieties, too many contradictory answers. So, as the twenty-first century begins, we might reconsider whom to trust and whose guidance to follow. Who might allow us to be fair minded, reasonable, and responsible in our responses to the increased pressures of technoscience?

I should hasten to warn that a solitary or individualistic response to the human predicament might not work as a prescription for a whole culture. Agreeing to "just do what you wish or follow your conscience" remains too vague, an ineffective way of conducting life. I might be an Enlightenment fool after all, but I do nonetheless suggest that what might work for one person could work for many. If we find a key to unlock the secrets of a tranquil life, then it should be a key that can be used by more than just a few privileged individuals. Others may require training and patience, but to suppose a priori that only a few can reach tranquillity is too painful an answer to the quest of the multitude. For example, the impact of the development and use of the atomic bomb has become a political instrument of global warfare, no longer limited to the scientific community. If the technoscientific community were willing to bear some responsibility for its scientific work, then the pristine environment in which they work, a sanctuary of sorts, would disappear and be open to public scrutiny.

I consider the synergy felt by those involved in "big science," a term used since World War II in the United States to describe large efforts of technical developments in diverse areas from cellular research to dams, to be explicable in terms of the ongoing quest for knowledge (and control) of the world. This quest is no longer linked to the limited theoretical development of the theory of relativity or quantum mechanics but is rather linked to the collaboration of large institutions committing numerous scientists to map out an ordered universe, to find answers to questions that bother the culture as a whole. The intrinsic value of scientific knowledge was transformed at the dawn of the twenty-first century into the extrinsic value of the almighty dollar. No longer do we talk exclusively of technoscience in the service of humanity but of how this or that corporation could patent and market an invention. To me, this transformation, which I might add does not infect every aspect of our culture, illustrates our personal anxiety and some of the methods we use to cope with it. Financial rewards have a meaning within capitalism that transcends the meaning of money alone; the meaning has to do with power, control, and order. The power to control part of our existence and establish an ordered environment for ourselves is a way to cope with an inherent sense of chaos and futility.

Two of the defining moments of twentieth-century technoscience in terms of utter confusion and the incomprehensibility of the human condition were the world wars. Mass destruction and the use of warfare as an expression of nationalism, power, and pride, whether focused on the atomic bomb or other methods of destruction and cruelty, changed our perception not only of the conquest of the universe but also of the hazards associated with that conquest. If we looked up to science to solve our problems and improve the conditions of human existence, we found out that other options were available as well. Scientists could no longer capture the image of high priests extolling the virtues of heaven on earth and acting responsibly on behalf of their culture; they turned into willing instruments of destruction.

The subservient answers they gave to the military and political authorities undermined the esteem with which they were perceived by the public at large. How can it be that the geniuses in our midst are also villains? How can it be that the methods of technoscientific development produce such horrible results? Public confusion led to anxiety, and public dismay led to anguish: What can be done? What can *I* do?

Learning the languages of nature, as we have during our entire history, is quite different from engaging in verbal (and unfortunately lethal) battles. The promises of technoscience in the past century have turned into multiple blunders; some we accepted favorably and enjoyed, such as health care, and from some, like nuclear fission, we suffered bitterly. Perhaps one way of concluding this chapter is to come back to the description of the human condition by existentialists. They are rightfully concerned about the futility of our condition, the absurdity of some of our efforts, and the sadness of our situation. But we should not settle for simple descriptions. We should find a way out of our miserable condition (as they describe it) and provide useful prescriptions. The way out might revert back to the stoics to the extent that we distinguish between those aspects of our condition that we have control over and those over which we don't. This way of thinking and living could encourage us to focus on those facets of our lives under our control. This can't mean only changing our perception of our condition but must also include taking actions that illustrate how we can hold each other accountable for our deeds. For example, we can more deliberately involve the public in decision-making processes that effect daily existence and that would turn all of us into active participants in our lives rather than confused or disgruntled observers. We can yearn for freedom and appreciate the compromises we must make to achieve it. We can yearn for certainty and appreciate the context under which we know things for sure. Finally, we can yearn for order and appreciate how open-ended order must remain. In each of these cases we can recognize that our compromises are deliberate and controlled, not signs of weakness and futility. Perhaps our change of mind will contain a change of heart, and perhaps this kind of change will lessen the sense of anguish we have, because we will be in a position to explain our actions in a manner that reflects responsibility.

CHAPTER FIVE

~

The Quest for Order (or, the False Hopes of Technoscience)

The previous chapter discussed the ambiguity, anxiety, and anguish caused by technoscientific advances that we carry with us into the twenty-first century. Science and technology have become the center of all debates, whether about warfare, politics, health care, or education. What couldn't be measured and mass produced has been considered meaningless by the industrial enterprise. What is readily available and could solve specific problems—be it missile guidance or tooth decay—has been considered positive and worth pursuing. Just as it became clear that our culture was infused with and informed by science and technology (as numerous writers have summarized the twentieth century, e.g., Prigogine and Stengers 1984), our culture has realized some unintended consequences that are bothersome or outright dangerous. We have learned to appreciate the problems with nuclear waste, and the potential for a nuclear holocaust, because of disasters, such as Three Mile Island in the United States, Bohpal in India, and Chernobyl in Russia. On a smaller scale, we have learned to worry about testing educational performances and the intrusion into our privacy in the computerized information age. Still, we are more likely to celebrate technoscientific feats without a critical eye and tongue, without the warnings that even spray paint cans must have on their labels.

The useful, even life-saving gifts of technoscience sold the public on new scientific ideas, such as instant global communication, voice-activated equipment for the handicapped, or mammograms to detect breast cancer. But, like Prometheus's gift of fire, technoscientific gifts can exact a dear price that requires a warning, if not a critical assessment. Some are so unpredictable and hazardous, like tampering with human genes and cloning, that they might better not be used at all. Mary Shelley's *Frankenstein* (1735) was just such a warning to an age that celebrated the advances of science and the hope of human control over nature. The monster Dr. Frankenstein created exemplifies the ingenuity of humans and the power of medical science. But once created, was the giant a friend or a foe? Did he obey his creator? Was he part of humanity or a machine with a human face? Does he exemplify the height of human creativity or the depth of its folly of trying to be godlike? These questions come to revisit us not only when we develop organ-sustaining machines but also when we manipulate genes in therapeutically beneficial ways, such as stem-cell research. Are we at the dawn of a monstrous age because of potential errors by researchers or potential abuse by politicians? Will Mary Shelley's nightmares become our reality?

Just as our conception of freedom raised expectations about our personal situations in life, so the promises of science and technology raised expectations about daily life. Just as our myths of freedom have transformed our culture over the past two hundred years, so have our myths about technoscience (as the cure-all of the future). We must clearly understand how to reconsider the dangers associated with technoscience in order to appreciate how they have influenced our personal views and attitudes. This consideration might help us realize how much of our own thinking and even feeling is bound to what happens in the culture that surrounds us. If our culture seems confusing and frightening, then it stands to reason that we would internalize its ambiguities in a similar manner. If our culture seems prosperous and friendly, then our own attitudes will be, generally speaking, positive.

Though I believe that culture, ideas and myths, customs and habits, inform our individual thought and behavior (in the Weberian sense of influence), I wish to avoid a crude or naïve reductionism. By this I mean that the influence of a theory or a set of values can transform the actual workings of a society, rather than the Marxist view that a set of material conditions influence the way we think. That is, I don't believe we are mere pawns in the games played by those dictating our cultural icons and images, those who set up the images that we are supposed to agree with as representing our desires and goals. I don't think there is a direct correlation between a cultural icon and an individual's thought. Numerous additional influences and subtle messages are filtered between a cultural idea or image and its expression by individual members of that culture. There is an interplay between our culture and our behavior; we are influenced by commercials and television programs just as much as we influence (in the aggregate of our consumption) what products our culture will endorse and promote, produce and distribute. The material conditions of our culture, as Marxists remind us, dictate certain limits and boundaries on our choices. But we have the power to adopt or reject some of these boundaries, challenge what is given to us, and reconsider what we are ready to put up with. We can and indeed do break with tradition when it becomes oppressive and cruel, as feminists have pointed out over the past four decades, and as we have seen with the fall of the Berlin Wall, and insist on change. It is in this sense that there is more of a reciprocal relationship between the individual and the culture, than a one-sided reductionist influence going in one direction only.

As mentioned in the first chapter, Epictetus warned us some two thousand years ago that we should recognize what is and what is not under our control. We should find a way to deal with those things in the world that are not within our control and focus more on those things that are within our control. Somewhere in the rush of technoscience, humans have forgotten that there are limits to our knowledge and to the control we can exert over the environment. In fighting our limitations, or what some call the human condition, we are like Don Quixote attacking windmills thinking they were giants. It is understandable that Don Quixote believed in what he did, just as it is understandable that we believe we can fight every disease and every natural shortage in the world. It's about finding order in the world, finding a way to classify, organize, define, and name all those variables in our lives that otherwise would remain confusing and mystical. This urge is reasonable only if we hope to accomplish something, get closer to our goal. But what is our goal, after all? Isn't it to live life with tranquility or peace of mind, as Spinoza called it? If indeed this is the case, then we ought to focus on what it would take in this new century to avert our attention from the confusions of yesteryear and find reasonable and useful answers to our quest.

The Cultural Basis of Individual Choices

What I'm talking about works simultaneously on the personal and the cultural levels. What society feels collectively, we feel personally. Not knowing if globalization is good or bad is similar to not knowing if marriage is good or bad. There are just as many arguments for as there are against opening national borders to foreign trade and immigrant labor flow. Likewise, there are just as many arguments for getting married or divorced as there are against. The old answers of institutionalized religions regarding why we should get married and shouldn't get divorced don't work as well anymore, because they have proved too rigid and often too authoritarian to stomach. Now, as an example of a cultural shift, single motherhood is culturally accepted, and the need for procreation in the traditional sense has been replaced with a fear of overpopulation. Church attendance is on the rise, and though religious affiliation in most of the major religions is increasing, religious answers to contemporary questions about environmental issues as well as health care remain wanting. If that were not the case, then the concerns of religious people and atheists alike would have all been laid to rest. But we as a culture and a community of interdependent people remain restless and confused about our political leadership, our financial well-being, and the future of our children and grandchildren. On some deep level, though many of us pay lip service to the church and respect its attempts to save lives and souls, some of us believe it doesn't have the answers we seek; we suspect that its answers have fallen short when crises occur.

Perhaps one of the most often cited examples of such a situation occurred during the Holocaust, when six million Jews were systematically exterminated alongside homosexuals, gypsies, and communists. The genocides of the last century should challenge anyone who still believes in the eradication of evil and the protection of a divine power. The Jews prayed and wept, begged and pleaded to be saved by their God, and nothing happened. The Nazi machine went on for years and was almost successful in eliminating an entire culture. I'm sure we can recall similar situation in regards to other people in our Western history (such as the Kurds on the Turkish border). The problem of evil hasn't been solved or explained away, regardless of the attempt to bring mystery into the fore, or to claim, as God did with Job in the biblical story, that humans cannot fully comprehend the universe in which they live and the mysteries of its design.

When Job was stripped of all of his belongings and his family was destroyed in front of his eyes because of Satan and God' wager to test Job's loyalty and faith, Job eventually became frustrated enough to challenge God. In one of the more memorable passages in the Bible, God confronts Job and asks him whether he was there during the creation, whether he, as a mere mortal, could fathom the design of the universe and the wisdom behind that design. Job ends up accepting what God says and is humbled before the voice and power of the Lord. Yet, I'm not sure readers today would take Job's position. They would be angry and skeptical when faced with losing everything they worked so hard to acquire and accomplish in contemporary culture if they saw no good reason for their loss. Maybe unforeseen natural disasters, such as earthquakes and hurricanes would be deemed acceptable, in some sense of the term, since their random occurrence wouldn't have targeted specific individuals. They would see a Job-like experience as complete injustice and incomprehensible suffering. Jewish survivors of the Holocaust remain just as confused as ever, demanding to know where their God was when they needed Him. To say that God is not responsible for the actions of other humans and that He doesn't wish to interfere and stop the cruelties of humans is a poor excuse for inaction (in the name of free will) and probably cannot make up for the mass murder of innocent children. But humans haven't given up asking questions and searching for answers, whether they have consulted oracles or prophets, shamans or soothsayers. Their continuous quest is what this book tries to explore. In this book, I try to explain the origins and history of this quest and to suggest that at the dawn of a new century the quest for order may be less associated with the (false) promises of technoscience than in the previous century. Technoscience, as we have already seen, remains ambiguous in its assessments and predictions, and it can only provide provisional or conflicting answers to the questions posed by an anxious public. The unattainable quest for order leads me to appreciate the extent to which we must act responsibly despite our inability to control the conditions of our lives.

It should be noted, though, that there are various interpretations of the meaning of order. For some, like Aristotle, the notion of order meant the classification of the world. Eventually this meant zoology, botany, and many other disciplines in which pieces of the universe are defined, then categorized with other pieces that are similar, and finally are distinguished from those that are dissimilar. Classification, of course, is highly problematic, for one's definitions may turn out to be incorrect or arbitrary, informed by social factors and connections and not by so-called pure science. When we get to the human sciences, the sensitivities of the last century have almost made the world impossible to classify at all. The French philosopher Michel Foucault chronicled some of these problems in his attempt to examine all the sciences as they were construed by the late twentieth century (1970). In his survey of the historical record, he tried to outline the shift from a confused view of the world to one where every aspect of our knowledge was organized, classified, and ordered. With the shift toward a more regulated worldview, we have become accustomed to monitoring our progress and our achievements in controlling our environment. Our ability to monitor our surroundings and ourselves establishes a sense of measured discipline that in turn regulates our lives. This allows us, in a scientific manner, to set models that explain what happens to us and that predict what will happen in the future.

Imposing order on the world and on ourselves is not necessarily the answer to our fear of confusion and unpredictability. Even when it appears to calm our anxiety, the order we impose on the world might be open to a reconfiguration of that order at a later time or to a hierarchical way of classification. Once a hierarchy is imposed, it leads to power relations that might turn out to be both cumbersome and discriminatory in the worst sense of the word. We end up ascribing a normative value to the hierarchy and its constitutive members, so that people might believe that something is better than something else just because of its place in the hierarchy (see Sassower and Ogaz 1991). For example, is mathematics or physics better than sociology or psychology just because they are considered more scientific or more accurate in their predictions? This question isn't as esoteric as it may sound at first, because it relates to the process by which hierarchies are established. It could as easily be applied to a hierarchy of the races, as the Nazis did.

Imposing Order on the World

Zigmunt Bauman has written eloquently about the rationality of classification and gardening, and the dangers associated with them (1989). For example, pulling weeds that hamper the growth of beautiful flowers could be seen as a metaphor for what the Germans did during World War II. If they are indeed weeds, why not eliminate the Jews, gypsies, and homosexuals from the German landscape so as to purify the Aryan race? As you can see, the definitions we impose on the world make all the difference. Physicians and psychotherapists are fond of reclassifying the list of diseases they use in order to cure their patients. Sometimes their changes in classifying diseases depend on new scientific discoveries, sometimes on years of experience. Odd situations arise, as in 1973, when the American Psychological Association voted in its annual convention to remove homosexuality from its previous classification as deviant behavior that warrants treatment.

Perhaps these stories tell us that the imposition of order on the universe and its inhabitants may be useful and efficient most of the time, but it can get us into a mental and emotional mess at other times. Perhaps we should instead focus more on an internal order to bring us peace of mind. I must hasten to add here that I don't mean to encourage a subjective and individualistic ordering that ignores reality, making us so reflective that we become insensitive to the community around us. Rather, I wish to emphasize how precarious and humanly informed ordering is. No matter what appeal we make to technoscience and the truths of our knowledge bases, we must remember that humans construct the world in ways that help them cope with their environment. As such, they have the power to confirm and refute whatever information they come across, reconstruct reality in a way that makes sense to them, and reconfigure how to live their lives.

For some people, it might simply mean organizing their daily routines and their way of making decisions in life. Some people call this "prioritizing," some "compartmentalizing." Setting priorities in your daily routines, whether for exercise or for reading bedtime stories to your child, is something we are familiar with. Sometimes it takes a crisis to make us aware of our priorities, when our health fails or our child acts up, seeking attention; sometimes a friend or therapist points it out in a conversation or consultation. Prioritizing is a form of ordering, of setting up boundaries and methods of interacting that express an internal ordering process as well as express what we believe to be important in our lives. In this way, we carry out what we have been thinking about. Incidentally, if you externalize without internalizing, the priorities will fall apart very quickly, because the internal foundation will be lacking (for example, when you pretend to care about the elderly but don't check up on your sick grandparents). Compartmentalizing, setting aside issues that can be discarded or dealt with later, is also a highly controlled method of making order out of a jumble of issues that might otherwise paralyze someone (because sometimes there is so much to do that you don't know where to begin).

If we follow the stoics, we might realize that even though we can't change the environment in which we live, we have the power to organize our lives to deal with it in a measured and responsible way. This is not about imposing order on the world; it is a modest suggestion to find ways of interacting with the world that make sense to us, that we can live with, that will bring to us some peace of mind. We have the power of mind to think through problems, set priorities, and act accordingly. Perhaps this is an ancient conviction regarding the power of the mind to influence our emotions and judgments of our surroundings (as I have outlined in chapter 2). The appeal, then, is an appeal to the mind, the brain, our rational side, rather than an appeal to our emotional side, our feelings and reactions to the world.

It seems to me that there is an intimate relationship between our reasoning capacity and our capacity to feel and react emotionally to ourselves and our surroundings. There is no priority or hierarchy in any strict sense between these two facets of our lives. Since there is a reciprocal and complex interaction between our minds and our hearts, so to speak, and since I have no keys with which to demystify our emotions, let me focus on the mind. For me, this means finding ways to discuss, convince, argue, debate, explain, and understand the goal of this analysis, and how we may achieve this goal. Incidentally, this might be called "cognitive therapy." This might be construed as a belief that we can do something about the futility of the world, the absurdity of our existence, and the meaninglessness of life. It is therefore a way of inserting ourselves into the world, so to speak, or discovering some meaning in what we think and do (in Frankl's sense). Perhaps it is a way of organizing ourselves according to some reasonable and explicit criteria that articulate our priorities, what values we adopt, and how we can reach peace of mind (tranquility).

Someone may ask: How should I order my life? How should I organize my daily routines? We don't ask these questions in a vacuum. Rather, we ask them in relation to our specific conditions, our specific circumstances. Put differently, we ask questions about our lives in relation to, and as a response to, a set of conditions imposed on us by someone else. The "someone else" can be our immediate family as well as the political and economic order of the country where we live. So, however limited and personal our questions are, however confined our quest for order may seem to us, it is undertaken within a much broader context. Our particular misfortunes are by definition related to, if not directly caused by, someone or something else. For example, we live in a run-down rented apartment not because this is our best choice but because we have no other choice given our education, profession, and the job market. Basically, this is the case because we don't have or earn enough money to live in a better place or own our home. The economic circumstances of our upbringing are beyond our control; what we can control are our reactions to and engagement with these circumstances—or as we say, make the best out of the situation.

The False Hopes of Ordering Life

The previous century brought with it the hopes of progress and success, the improvement of the human condition at least medically and economically. By this I mean the ability to explain some absurdities away and to find worthwhile goals to fulfill so that life doesn't seem futile. The fact that the promises and hopes were not shared equally or that their fulfillment was not as widely available to everyone brought with it confusion and disappointment. This reaction, as it spread over a growing population, provoked a sense that the world was not as well ordered as we'd hoped, that there is more chaos than order when it comes to large populations and communities. Despite a robust economy in the United States at the turn of the century, we experienced the Great Depression and the collapse of the stock market in the 1920s. Business cycles can be explained retroactively, but economists haven't yet figured out a way to prevent them from taking place altogether. Similarly, we are proud of living in a model democracy, but we acknowledge how little power the general public has over the institution and implementation of public policies (from taxation to gun control). Has the world become more explicable and predictable in the past century, or are we faced with utter bewilderment when facing our future? Those paying into the Social Security system at the end of the twentieth century were less confident than had been parents that the benefits would indeed be paid out when they retire.

Suspicion and skepticism replaced hope and trust, and before long the Western world as we know it turned into a puzzling labyrinth. Who are our enemies? Who are our allies? The menace of the "evil empire," the former Soviet Union or Iraq, for that matter, is no longer threatening our own country. Do we have enough energy to run our economy, or will some foreign supplier cause an energy crisis like the one we experienced in the 1970s? Will we find our way out of these situations? Can we make sense of what is happening to us, to the world around us? Senseless wars ensued, innocent lives were lost, and our entire worldview seemed to fall apart before our eyes. What can we count on? Whom can we count on to rescue us or to provide meaning in this chaotic world? Religious and cult leaders came forward and offered their services, some more cheaply, some more dearly than people expected. Academic institutions offered higher learning as a way to find answers, if not financial rewards. The culture as a whole tried to find peace of mind through wealth, saying, in effect, that if one had enough money, she or he could buy enough material comfort to become insulated from the surrounding confusion of the world. You can be an island onto yourself, have moats to protect you, and live in relative security in terms of the structure and order installed in your own life (as gated communities advertise their security systems and their isolation from the outside world). You can watch the rest of world on television and leave its confusions and concerns to the front pages of newsprint. It's really not about you!

What I provide here is not a "how to" manual that outlines how everyone should live. Rather, I want to illustrate here that our personal feelings about establishing order in our lives are informed by a culture that, with its claims of freedom, has taken away from us tradition, custom, and habit. We are all alone with the power to do "anything," to "be all that we can be," and before we know it, we have to make responsible decisions about everything all the time. Wouldn't it be easier if we knew what our lot in life was? If we knew, as Socrates tells us in the *Republic*, to what class we belonged and how we were designated to contribute to the whole community, we would be secure in our social roles and content with our position. Oh, yes, we could rebel and move upward or sideways, change our predestined profession because we exhibited some extraordinary talents nobody foresaw. But this would be the exception, not the rule. A youth in Western contemporary culture is pushed to decide what the future will hold; the options are immense, and the intimidation in the face of so many options is daunting as well. So, what is one to do?

I guess this is where my concern for navigating through the ocean of confusion comes into play. This is where the twenty-first century is heading, finding ways to use our navigational instruments and hone our skills. It's a century that will be ordered, organized, codified, and compartmentalized so that we don't wander toward the abyss and fall off the edge. The existentialists of the past century suggested that we find something we are passionate about, something we care deeply about, so that we can divert our attention from the absurdity and futility of life in general and pursue a personal goal with conviction and commitment. But in order to find our path and realize what brings us peace of mind, we must appreciate and understand the context, the environment in which our choices are made. I'm talking about the social and cultural environment, rather than the natural one. In order to realize this environment, I shall examine some of the more prominent institutions that dominate the cultural scene in Western societies.

One of the oldest institutions is religion, whether one follows the Judeo-Christian tradition, as so many are fond of calling it, or other religious doctrines. The beauty of this institution is that it has a sacred text, the Bible, around which all rules and regulations, all ceremonies and restrictions are based. Commentary on the Bible in the past two thousand years has primarily focused on two concerns: finding ways of reconciling apparent internal contradictions and applying general rules to the social and political environment of the day. At times, the application became itself a new guide for behavior and turned into a habit or a custom, garnering the status of a rule (though it was only an attempt to apply the original biblical rule). This was the case with *Kashrut* (Jewish dietary laws), for example, even though contemporary refrigeration techniques could dictate a different interpretation of what could and could not be safely eaten today.

The Bible provides a set of laws, such as the Ten Commandments, and a series of some 613 duties that must be fulfilled by observant Jews. The good news is that if they are followed, one is considered righteous and can expect rewards from God. If one fails, the consequences are numerous, but all of them are somehow outlined in one form or another. If one has a quandary, question, or puzzlement about any aspect of life, there is always a text or a rabbi to provide precise and concise answers. An order and responsible action is rendered from within the institutions of religion with the sanction of God, a divine entity whose authority is beyond reproach. The attraction to religious institutions is immense and justified, if my thesis about the human condition at the end of twentieth century is correct. If indeed we are frightened by the confusion and chaos that have been handed to us by our culture in the past century, then we would gladly adopt any measure to soften the blows of confusion and find true, permanent, and reassuring answers with divine sanction. The appeal of religious leaders and their pronouncements is almost unparalleled by secular institutions.

What secular institutions, such as the military and government agencies, have going for them is a detailed set of rules and regulations, all the way down to the way people dress, eat, and sleep. Not much is left for individual choice, not much to wonder about. From the moment you wake up to the moment you fall asleep every aspect of your life is dictated by someone else, leaving no room for confusion or for heart-wrenching decision making. Choices are made for you, and you can leave the institution if you object to any of them. Just as religion provides a whole range of rules to follow, so do these institutions. In many ways, these institutions simplify life to the extreme, reducing the ambiguity, anxiety, and anguish talked about in the previous chapter. If ambiguities are removed, because you know exactly what is expected of you and what your rewards or punishments will be, then there is no anxiety about what course of action to follow. In the same vein, once you have no anxiety about your next move, you are less likely to have anguish about your life as a whole. All the pieces of the puzzle fit so perfectly that putting the puzzle together becomes crystal clear. Every minute and hour is accounted for, and life as a whole makes perfect sense.

Michel Foucault is one of the most brilliant chroniclers of the prison and military systems, institutions where discipline is highly valued and adhered to in every facet of life (1979). Perhaps he chose those institutions where the body as well as the mind is disciplined to illustrate how confined our lives have become. Perhaps he wanted to warn us against having our lives so regimented that we would begin thinking and living like soldiers or prisoners. Socrates warned us that the unexamined life is not worth living, and Foucault seems to continue in this vein. But examination is difficult, even painful. It requires considering alternative options, having to choose, and then discovering that you may have made the wrong choice. It means enrolling at a university while worrying if higher education is indeed fulfilling, enlightening, and worth the effort. The soldier and prisoner alike find some comfort in having someone else make the choices, and then being able to complain, even protest. The congregant likewise prefers to follow the priest who claims to speak on behalf of God, providing exact directions on how to live life and what to do at every stage of life.

It is not surprising that in this cultural environment (where options breed anxiety about the future) many of us prefer to join institutions where there are specific guidelines of behavior, codes of conduct. These institutions need not be the military or a police force; they can be a university or a country club. However more relaxed in their guidelines they might seem at first sight, they turn out to have detailed instructions about the expectations of one's behavior, so that a sense of order is brought to everyone's attention. Students have rules of conduct and peer pressure to guide their choices; country clubs mandate what you wear and how you play (golf, tennis, or anything else) with your fellow members. If you don't feel comfortable with an order you impose on yourself, then you might welcome conforming to the order imposed by others, especially if they intimate rather than demand it. You, the individual, don't have to figure out things alone; they are preestablished. Just follow the rules, and you'll reap all the rewards associated with and promised by the institution to which you belong. Deviant behavior is not tolerated, and you may lose the tranquility of your membership, your sense of belonging, if you fail to abide by the rules.

Technoscientific Gifts and Their Price

As understood in the last century, technological innovations might have unforeseen side effects, changing more than initially intended. In short, as the political theorist Neil Postman argues, technology brings about an ecological sense, to the extent that it transforms the whole environment (1992). When we introduced the automobile, we didn't just provide a quicker mode of transportation. Rather, we added roads and gas stations, we refined oil and developed living spaces far away from factories. The automobile changed the way we live, not only the way we move from point to point. To think of technoscience in the past century is not only to think of gadgets and specific breakthroughs but to realize how much we have transformed everything about ourselves. Our health and life expectancy have radically changed, our clothes and homes only remotely resemble those of our grandparents, and the way we plan our future—with aspirations and dreams we believe will come true—is completely different from the way our ancestors thought of their life plans.

Whether we focus on the scientific revolutions of the sixteenth and seventeenth centuries or the ideals of the Enlightenment, it's clear that technoscience introduced the expectations of human progress and happiness. Unlike the promises of religion, the promises of technoscience could be realized here on earth within a short period of time. Both institutions have a mystique about them; they use languages not commonly used among those engaged in daily commerce, what we call "common folk." They have rituals and uniforms that separate them from the rest of society, white coats, long black gowns, academic or theological degrees. They have set themselves apart so as to distinguish themselves from everyone else, to draw attention to their authority and the power they can exert on the rest of society. What's most astonishing about their success, perhaps the key to their success, is their presumption that their views, images, and ideas are not limited to themselves but are universal. That is, just as religious leaders have offered answers that are eternal and all encompassing, so technoscientific leaders offer views of the world and its future as comprehensive, true, and beyond doubt. For example, the church tells us why we might suffer eternal damnation, while medicine tells us why we may be doomed to die of cancer. Both claim certitude, both appeal to the Truth, and both have us frightened to death of their verdicts. Moreover, both judge our behavior and our actions.

Remember that it was doubt that motivated our modern thinkers (since Descartes in the 1600s) to establish immutable truths and the sense that once a scientific truth was proven, it was proven beyond the shadow of doubt. Certainty—knowing for sure, 100 percent—was eventually transformed into probability—knowing what is likely to happen, in most cases, most of the time, to most of the people affected by a situation. But even those of us who appreciate the subtle shift from certainty to probability still revert to certainty, because it offers more conceptual stability and security than the ambiguities opened by probability. Most of us tend to forget that yesterday's certainty became today's doubt and turned into tomorrow's folly. We shifted from a geocentric universe (the earth is at the center) to a heliocentric one (the sun is at the center), only to agree with Einstein that the very notion of the center of the universe doesn't make sense (there is no center as such, since all planets are positioned relative to each other's fields of gravity). We became scared of this nebulous answer and tried to transform it into something more stable and secure. The unpredictability of the world—as the Russian scientist Prigogine already suggested (1984) and the American journalist Gleick eventually recorded (1987)—was thought to be chaos. Chaos became more ordered over time, with paths and patterns, so that some order could be discerned. The so-called theory of chaos is a model that is explicable and predictable in principle. It's a model that only traces the antecedents of chaos but in effect insists on order and regularity, stability and certainty (in a weak sense of the term), so that we can comprehend it, deal with it, and own it, so to speak.

Regardless of the advances of the twentieth century and the fact that the public is much more widely educated and better read (having almost erased illiteracy in the Western world within the past century), the world still seems incomprehensible to most of us. We seek signs and guides, answers that make sense, and assurances that our knowledge will help us understand whatever confronts us. When we fail in our relationships with lovers and friends, family members and employers, we seek the assurances that the external world, the sun and the moon, the rains and the droughts, are within our cognitive reach. So, even if we can't figure out the irrational outburst of a friend when meeting at a restaurant, we can still find comfort in the empirical (expected) fact that the car we parked outside the restaurant will be there when we leave. If during our process of education and maturation we have to rely on technoscientists rather than priests, it is because we believe their answers are better, more useful, even closer to the truth. Some, like Foucault, find the institutions of order oppressive because their concern with discipline undermines our ability to examine and critically evaluate the data given to us. Others, like Postman, suggest that the establishment of technological institutions and a mind-set overwrought with technological images and fixes are attempts to control the overwhelming amounts of data given to us. The role of these institutions is to find ways to explain what the data could mean and ways of ordering them in intelligent manner. But can these institutions succeed? Isn't there something inherently problematic in their approach and pretense?

What may be inherently problematic with the approach to twentieth-century technoscience is that too much was expected from its success. In this way of thinking, the human condition itself should have improved. Society would be perfectly organized and human happiness could be guaranteed. One could (in principle if not in practice) outline the path taken in life and find a causal chain of reactions to explain why certain things happened the way they did. If this were possible, we would feel less miserable about our lot in life and feel that there is meaning in what we do and how we interact with others. This sounds awfully familiar, dating back to the utopian visions of Socrates and all those who followed him. Whether the approval comes from a philosopher or from God, what is at stake is the improvement of the human condition in terms of leading a meaningful, satisfying life. The difference between the historical rendition of utopian visions and more recent ones lies in contemporary culture's address of the appeal to human happiness—how this cashes out in our daily lives, since the appeal itself (that life can be meaningful and happy) has remained intact.

Technoscience has been a willing participant in this game, setting up expectations that are unrealistic and even dangerous. Will the computer age bring equality to the world and enhance democracy (equal access to all data)? Will it bring hope to the poor and content to the rich? Will this utopian appeal bring about new generations of egocentric, reclusive, and antisocial people who couldn't care less about their society? Or will it, by contrast, bring people together from all walks of life in a manner never envisioned before? The promise was one thing, the unintended consequence quite another. Perhaps the difference on this level lies in the personalization of the utopian vision, because the focus has become the individual rather than society as a whole. In truth, we as a society are less concerned with the sacrifices we might have to make to ensure the well-being of everyone else than with the benefits each individual must enjoy under democracy and capitalism. Ideas and visions are translated into material comforts, and the measurement of happiness is linked to how much we own rather than how we feel about what we own. The fetishism of consumption, as outlined by Marx and closer to us by Thorstein Veblen (1899), the appetite for satisfying all our desires, is being promoted by marketing companies that perpetuate a way of thinking and behaving commensurate with an ever-increasing appetite for consumption. How often have you heard someone telling you that you earn too much money or have too many toys? How often have you been warned against (rather than on behalf of) conspicuous consumption? Why not buy the most expensive car or home or dress? Why not show off how rich you are?

The Technoscientific Effects on the Human Condition

We should not blame the leaders of the technoscientific community or the industrial complex that has evolved around the tools and skills of that community. Perhaps the leadership finds itself in a role it didn't seek for itself; perhaps the anxiety of the public has propelled its members from technical obscurity to celebrity status. Perhaps, as they often claim, all they are doing is feeding the desires and needs of consumers who want to consume and enjoy more of their products as they bring them into the marketplace. When I examined elsewhere (1995) the culture of expertise that has dominated our society in the past century and a half, I found experts and their professionalism blameworthy for the public expectation and for the rewards they could reap from their newly acquired status. But when examining the culture of technoscience during and after World War II, I put some of the blame on a confused public that put its trust and pinned its hopes on the ever-increasing fruits of technoscience (1997). The public seemed to have relinquished its responsibility for finding answers to pressing questions, expecting others, technoscientific experts and leaders, to solve all the problems of the world without a hitch. By this I mean not only the problems associated with, say, pollution but also with physical fitness and personal happiness—from nutritional advice to all sorts of therapies. The responsibility doesn't lie exclusively with one group or the other but is in fact a result of a reciprocal process, reminiscent of what psychoanalysts call "transference" and "countertransference." At the end of the day, it really doesn't matter who conferred the role of soothsayers and oracles on technoscientists. What matters is that technoscience cannot offer the gifts expected of it; it will always come short of providing universal happiness to an anxious and anguished public. Even when it's willing and able to satisfy its customers one day, inevitably it will fall short the next, because with more food the appetite increases.

Furthermore, the twentieth century also witnessed a split between culture as a whole and the culture of technoscience, as if the one has nothing to do with the other or is alienated from it. Of course, this is not true. The British thinker C. P. Snow complained in 1960 about the two cultures, humanities and science, and Alexander Koyre has alerted us to the two worlds set up by the scientific discoveries of the past three centuries (1968). Even if the mysteries of the heavens have been turned into explicable and predictable models, the world as we know it remains full of enigmas and mysteries. We know ourselves better in some scientific sense (what causes a specific ailment or what relieves a particular pain), but we still lack a deep understanding of the meaning of our lives. We are still mired in the situation described by the existentialists of the past century (Camus, Sartre, de Bouvoir). Quantitative analyses, measurements of sociopathologies and psychological anxieties of individuals and classes of people, are useful and informative. But their limits are made clear when qualitative transformation does or does not come about. We see no gestalt shift in our conception of human destiny or the way we acknowledge another person we meet on the street. A good example of this predicament can be observed daily in regard to health care, where the advancement of medical science and its attendant technologies has left us as bewildered as we ever were in regard to the quality of life and the wisdom of dying with dignity. Medical staff can pump medication into your veins and hook your organs up to a machine that will sustain them for a while, but what if you don't recognize your relatives or cannot make your wishes known to the attending staff at the hospital?

No matter how accurate our quantitative analyses are, no matter how powerful our technological tools to quantify and calculate, we are still bound by our own perceptions, traditions, and customs. We can go only so far in the development of our environment before we find our own limits, and the limits of our imagination (see Pool 1997). As the experiences of World War II taught us, the technologies that were at our disposal have outstripped and overshadowed our own visions and conceptions. Whereas it has been standard procedure to imagine the future, anticipating what we could do if we only had adequate tools with which to accomplish our wishes, the experiences of the past century taught us otherwise. We couldn't have imagined human cruelty in gas chambers, using efficient measures of human destruction and Zyklon B as the chemical of choice. We couldn't have anticipated the use of engineering precision regarding transportation to be used for mobilizing millions to labor camps and extermination sites. Likewise, we couldn't have imagined in our wildest dreams and nightmares what effect an atomic bomb would have on civilians in Hiroshima and Nagasaki. Technoscience carried us with it, not as if it were out of control (in Langdon Winner's sense, 1977) but rather as if it were unfolding new horizons we had never fathomed before. There is something exciting and frightening about this way of doing business, because it reverses what we traditionally considered the human course of action. Instead of one's having an idea that might eventually be implemented with the development of technoscience, technoscientific feats posed novel ideas we hadn't anticipated. Our imagination followed our tools rather than our tools being developed to catch up with our imagination. (For more on this, see my *Technoscientific Angst* 1997.)

The excitement of a technoscientifically sophisticated culture comes from being surprised and learning new things that were not thought of before. Who would have imagined hand-held cellular phones that communicate data from remote areas? The fear comes from having unanticipated, devastating consequences; worse, these are consequences that already happen and cannot be stopped, such as shortened time intervals in which we must respond to queries or transmit our decisions (as with Internet electronic mail). The arrow of time, as we have learned to accept, is faster than ever before. What's more, it's irreversible. We can't stop the clock or undo that which has been done. The great technoscientific progress of the twentieth century seems in retrospect less promising as time goes by—progress in technoscience but regress in the human condition. It seems as if we have less time to think and reflect on what we should do, because we are expected to react immediately to others. The constant pressure on our cognitive ability has turned us into mechanical devices that use routines and fast responses, that lose the sense of humanity we have taken for granted for so long. We seem to have become accustomed to this way of life, rather than be bewildered by it.

Jean-Jacques Rousseau had a similar complaint about the progress promised by the Enlightenment, recalling his sense of the "noble savage" who at least displayed some pity and care in his interactions. Are our enlightened and well-trained lawyers and brokers on Wall Street friendly, concerned, and altruistic? Or have they instead used all of their education and skill (their presumed enlightenment) to pursue their ambition and greed at any cost? What has happened to our leaders as compassionate mentors, role models, and thoughtful counselors? These questions are not meant to take us back in time and portray a nostalgic past that never existed, since there always were unscrupulous, self-serving, and greedy leaders. Instead, these questions are meant to evoke an aura of care and concern among people, not despite technoscientific accomplishments but because of them. We need more careful analysis, not less. We need more caution and foresight, not less. Above all, we need prophets in our midst to warn us against our potential follies.

These questions may seem anachronistic, even a throwback to a romantic era we have long forgotten. But questions about the quality of our lives, our future, and the future of our children haunt us to the extent that we still seek answers. Perhaps one way we have learned to cope with them is to organize our environment as well as our worldview. Our quest for order, as I have argued all along, is both reasonable and attainable; it's something we believe would make our existence tolerable if not more enjoyable. It's a necessary coping mechanism in a changing world. Whether this quest can be satisfied with technoscience remains to be seen. How much we have learned through technoscience and how much we have been able to organize our lives with its aid remain indisputable. But is this the kind of order or the kind of coping mechanism we have been seeking? Do day-planners, alarm clocks, flight schedules, or global positioning systems bring about peace of mind? Yes, they help us avoid disasters and rescue individuals caught in storms and avalanches. But do these devices and instruments help us lead happier lives? These questions must be answered in this century in a manner different from in the previous century.

One way of thinking about this challenge is to focus again on the human condition in an existentialist sense and not in the sense of the conditions of humanity. This could mean paying less attention to measuring the material improvements we can enjoy and more attention to the ways in which these improvements affect our attitudes toward our fellow humans. Have we found answers to the question about the absurdity of human existence? Have we been able to insert meaning into our activities, thereby overcoming a sense of futility? With a fast-moving and ever-expanding cultural universe there is plenty to keep us busy—we can observe more than we can absorb or comprehend. Our observations keep us detached from what we see, so that we remain as alienated from our culture as ever before. We observed church rituals and were anxious to understand them and be forgiven for our sins, find salvation, and be rewarded in the afterlife. Has anything changed? We observe technoscientific rituals and are anxious to understand them and be forgiven for our ignorance, find (someone else's) meaning, and be rewarded in this life. This is commonly the experience of those who come in contact with medicine. We expect our medication and treatment to take effect right away, from the common cold to fertility. But can we fully insert ourselves and fully engage in the improvement of our lot?

It's clear that the innovations of technoscience and the conceptual framework that underlies them, while amazing in scope and effectiveness, do not serve to ease the anxiety of our Western culture. They seem to offer gifts, even a hope for constituting a more hospitable environment for our existence, but they do not suffice to assuage our anguish. It seems that the existentialists of the past century captured the sadness of the human condition, especially in light of World War II, and proposed tentative solutions. We must be reminded of their ideas, since they incorporated the reality of their time instead of ignoring it. When proposing an ethics of ambiguity, for example, de Beauvoir acknowledged the predicament of humanity. Likewise, when proposing hope and personal commitment, Camus recognized them as acts of conviction in the face of futility and meaningless. It is in their path that I follow with my proposals, realizing that technoscience is bound to unveil more problems than offer solutions, and that the human condition might not be transformed with changes in the conditions of humanity. The only price demanded of us is to be accountable for our actions, think through the reasons for our actions and their consequences, and uphold our share of the responsibility for whatever happens around us (from recycling to pollution, from energy consumption to instant communication).

CHAPTER SIX

~

A Modest Guide

Search for Meaning

To participate in the twenty-first century will require quite a bit from young people. On the one hand, it will require knowledge and understanding of the conceptual frameworks implicit in their culture—from the appeal to rationality to the expectation of personal freedom. On the other hand, it will require a level of personal commitment to changing the world in a responsible manner. In short, it will require people to be ethical. By "ethical" I don't mean a subscription to any particular religious doctrine or secular worldview. Rather, I mean the constant critical evaluation of one's actions and reactions in terms of a set of principles or values. The principles and values themselves can come under scrutiny and may be found wanting, but this is part of being ethical and not callous.

I offer some suggestions for how to embrace order in our daily lives in a responsible fashion and how to transform our view of the world into a more workable set of ideas. I propose to adopt a combination of Epictetus's and Frankl's ideas as they relate to the levels and ranges of things that humans can and cannot control. The control factor in our lives, in the sense of assigning meaning to what we do, affects how we reduce stress and appreciate the freedoms that we retain regardless of the oppression of technoscience. I also suggest that the confusion of the past century and the realization that our humanity is at stake (as shown in the previous chapter) might pave the way for an open discussion of our quest for order as a coping mechanism in a rapidly changing environment. In this sense, then, the goal of this book is to empower individuals to take charge of their own versions of the human condition without fear that a critical quest for order will lead them into the hands of totalitarianism. By taking charge, I mean infusing meaning into anything we do. By taking charge I also mean being accountable to ourselves and to others when we get dressed, when we perform a task, or when we communicate with others.

I should add here that by "the human condition" I mean the sense of futility described so carefully by the existentialists of the past century. From their perspective, it has been the morbid dread of death coupled with the absurdity of fighting to remain alive that has emptied all meaning from life. If you add Frankl's descriptions of the experiences of inmates in concentration camps, you can immediately appreciate how vulnerable humans are not only to the conditions of their existence but also to the cruelty of other humans who are more powerful than they are. The insanity and benign terror of gas chambers, the systematic annihilation of millions of innocent people, and what has come to be known as the "banality of evil" have all contributed to an intensified sense of loss and meaningless of life. As both Epictetus and Frankl implore us, it's up to every individual to search for meaning and expand energy on those aspects of life that are within the control of that individual. At times, this suggestion is overwhelming, as one can testify in the case of concentration camps. At times, this is a more mundane undertaking, when an individual commits to an idea and goal and makes an effort to find meaning in his or her activities.

I come back to technoscience and postmodernism, both of which are in fact realities we have learned to live through at the end of the past century and into this century. The botched presidential election in the United States in 2000 is a sad testimony to this fact. Instead of celebrating the ambiguity of political messages in the country and the ambivalence of the popular vote that gave a small margin to the candidate who eventually lost the election to another who was anointed by the U.S. Supreme Court, a postelection anxiety and anguish was expressed throughout the multimedia blitz that seemed incessant in its delivery of confusion and cultural anguish. The greatest power on the globe, the greatest democracy on earth (as Americans claim), came to its knees in front of television and Internet screens across the nation. What has happened? What can we do about this situation? How can we survive this chaos?

Order was restored by the Supreme Court. The law of the land, however politicized, took charge of our political lives, and the markets expressed dismay—the stock market plunged to decade-long lows. We want an orderly election, markets that keep rising in their value, an education system, a police force, and a military. We want predictability and security, explanations that make sense to all of us, and media that tell a narrative we can all comprehend. When our surroundings are confusing, when we cannot assume stability, we tend to revert to the worst of our human instincts of fear and greed, as one of our former secretaries of the treasury and financial leaders, Robert Rubin, declared in one of his speeches at the Colorado College in 2001. Are we back to Thomas Hobbes's sad descriptions of humanity as an ongoing war among all individuals in his *Leviathan* of 1651? Are we back to an era we thought we had overcome by the eighteenth-century Enlightenment? Are we facing the worst of nightmares of George Orwell's *1984*, where individuals feel alienated by Doublespeech (a language in which truth and falsehood can't be distinguished or ascertained)?

As a new millennium has dawned, we might turn our backs on the promises of globalization, even on national politics, and return to the communities in which we work and live. "Think globally, act locally" was a slogan from the 1980s that has been a helpful reminder of the limits we must draw in order to be politically effective. We find ourselves involved in our daily toil and focus more attention on those aspects of our lives over which we have control. Our votes count close to home, because they seem immune to the megabucks advertisement campaigns of national politics. Our trust in our neighbors is restored when we can directly exchange ideas and gifts with people we know, with people who will remain accountable to us because we live right next door. The inevitability of future interactions makes the present exchange necessarily meaningful, fair minded, and even just. Our expectations are higher of those who we call friends than of those we don't even know by name. This is not necessarily Schumacher's *Small Is Beautiful*, as to the extent of all economic development (1973), but at least a plea for localized social and political action. It's also a way to infuse our local interactions with meanings that we can define and defend, such as free speech and private-property rights.

Aristotle argued for the importance of friendship, both in interpersonal interactions and within communities, in his *Nicomachean Ethics*. Though the premise of his argument rested on a well-developed sense of virtue and character that ought to be displayed by all participants in a social and political organization, he allowed that those less qualified could be trained, educated, and indoctrinated to follow the role models of their age. Aristotle's premise was utopian in the same sense that Socrates' was; and theirs were ideals eventually adopted by any thinker who envisioned a utopian world in which peaceful coexistence would become a reality. Just as Aristotle conceived of his writings as guides and learning tools for generations to come, so did the Enlightenment leaders hope to promote the perfectibility of humanity through broad-based education. But have educational institutions delivered the goods? Has the educational establishment become the beacon of hope and transformation, or has it indeed retained power relations and class structure in such a fashion that it undermines said perfectibility? Put differently, has education had the perverse effect of undermining the improvement of morality and the building of virtue in every citizen?

Personal Education

As I have argued elsewhere (2000), educational ideals may have remained similar to those espoused some two hundred years ago, but their institutionalized setting has changed their tenor and practice. In an age of professionalization we are more concerned with the acquisition of skills than with the creative development of the mind. We are more concerned with teaching students the latest computer language or the underlying code that makes that language effective for their purposes than with the history of computer languages. This is an unfortunate turn of events, because it seems to me that a valuable lesson can be learned from studying the history of one's chosen field. For example, if you know what computer language or what code was replaced by another and for what reasons (too slow or too complicated to change), then you'd be in a better position to anticipate the shortcomings of whatever language you are currently using. This knowledge would be extremely useful in setting the stage for the development of the next language or the next code.

This, incidentally, is true not only of computer languages but of physics, chemistry, biology, and all the social sciences. With a heightened appreciation of the evolutionary nature of knowledge acquisition we could hope that our students would be both critical and creative. They would be better positioned intellectually and technically to figure out how and why theoretical dead ends were overcome in the past and how creative ideas came forth. If they could learn these lessons, they would also be better positioned to invent new methods and new tools with which to overcome obstacles and dead ends. In the competitive marketplace of the twenty-first century, an advantage like the one just outlined would be immensely important for companies the livelihoods of which depend on coming to the market with the latest, best, most efficient software product. So, what might seem at first sight an academic concern with learning the history of your chosen trade turns out to be a financial necessity. Investing in this allegedly idle study and contemplation is not a luxury, as some claim, but critical for success.

When my students study the history of ideas, they realize which problems have been solved and which remain perennial. They also learn how to think about contemporary problems, such as those that present themselves in the medical arena (comatose patients, organ transplantation, genetic treatment, and experimental therapies), in ways that have never been thought of before. They don't just use the skills and knowledge of the history of philosophy in their daily lives but also recognize what alternative solutions would or wouldn't be appropriate. For example, we can teach all day long the classical ethical models of past generations, but they would be deemed useless theoretical exercises if they weren't reformulated for the conditions of the present. Technoscience has changed the landscape, as already noted in previous chapters. It has even changed our human interactions and our sense of identity (with sex changes and other medical interventions). As such, the current situation requires new ethical models. But in order not to reinvent the wheel or repeat the mistakes of previous eras, we must know about the classical models of ethics as much as possible.

There is a personal level in which these lessons apply for college students, especially in the current era. The Internet, for example, provides large databases related to the various disciplines for which courses are offered. There are even services that provide ready-made papers on all kinds of topics (that might be required in standard course work). It's up to the individual student to make proper use of these tools rather than merely print out a paper and hand it as if it were the student's work. Whether we call it lying or plagiarism, there is an ethical code a student can either follow or trespass. Similarly, a student can communicate with professors and fellow students using electronic mail for the purpose of exchanging ideas and gathering information, asking questions and scheduling meetings. The same student can abuse these technological tools to harass other students, invade the privacy of the professor or fellow students, or alter university records. Computer technology can replace some of the uses of a library, but it must be treated respectfully and responsibly in order to maintain the kind of ethical standards we need in order to sustain our humanity and the institutional setting in which we study.

In short, students' and professors' behavior in the advanced age of technoscience should be in accordance with the same sort of guidelines and safeguards we have had for generations. The violation of any of these codes of conduct isn't justified because of the advent of new tools; however tempting these tools are for violation of old codes, they should be treated even more carefully than ever before and avoided altogether if the temptation is too strong. Likewise, I suggest that when those involved in the academic world enjoy the privilege of having at their fingertips the latest technology or the latest tools of technology, they should remember that this is a privilege the rest of the world might not be enjoying. This should make them both more critical about their own actions and more accountable to themselves and others. With privilege, as some of the aristocracy of yesteryear understood, there are duties (*noblesse oblige*). Those duties should not be taken lightly.

To be ethical in contemporary Western society also means to listen and learn from colleagues in a way that isn't limited to self-interest. By this I mean the critical exchange of ideas and thoughts, the process of listening and offering opinions in a way that is mutually respectful. A critical exchange can be accomplished without giving up the sincerity of disagreement but with an open mind to change one's views and agree with someone whom you might dislike on a personal level. Technological feats elicit awe as much as intimidation. They set up hierarchies of expertise that are at times so daunting that the novice, say, a student or a professor, might not even try to comprehend them. It's our mission and responsibility to make novel ideas accessible and sophisticated instrumentation user-friendly. Intimidation by force or rank, by authority and wealth, by divine revelation, can be replicated in contemporary culture when only a few have access to and know how to use the latest tools of technoscience. If we are to be different in the twenty-first century in our interactions with others, we must insist on egalitarian principles and highlight the promises of freedom and liberty. Those ideals can be either readily more or radically less achievable. Which way it goes depends wholly on those of us in position of power and authority, those of us who can make the difference, namely, those of us who are more educated than others and have more control of the means of cultural capital.

On yet another personal level, education is not limited to those engaged in institutional higher education. Television can be used for educational purposes, since there are enough channels that provide access to the arts (plays, movies, international newscasts, scientific data [Discovery Channel], and political debates). Likewise, we can read books and go to free public lectures and concerts on any given day. Of course, all of this requires a certain effort that is different from seeking entertainment alone, but even the so-called educational venues of our culture can become a form of entertainment if we put our minds to it. It's sad that we have forgotten the lessons of ancient Greek culture, which implored its members (admittedly only white males) to have a healthy mind in a healthy body. Have we skewed our attention to worrying only about a healthy body? Have we become obsessed with exercise machines, gyms, outdoors activities, and diets at the expense of exercising our minds? I tell my students and friends sometimes that if only they spent as much time on their minds as they do daily on their makeup and bodybuilding we would have a different society than we have today.

I should hasten to say that my recommendation isn't limited to reading the great books of our tradition or logical exercises. Any form of meditation can bring us a long way toward enlightenment about ourselves and our roles in life. Any sort of critical self-reflection can increase rather than decrease our happiness, even if we discover that there are some things we hate about ourselves, about our behavior, or about those with whom we associate. It is our realization that something needs to change that is the first step toward that change, toward an improvement already envisioned by the Enlightenment figures. If you refuse to recognize how your lot in life fits into the big picture, you will be at a loss in terms not only of why you are feeling bad in a particular situation but also of how to change your own predicament.

Another way of explaining the kind of advice outlined here is to suggest to people the quality of seeing the big picture while attending to all the details that make this picture what it is. This is no different from the workings of artists who must appreciate beforehand what the canvas should look like, while being forced to work on it one brush stroke at a time. This is true of business owners and operators, who must have an annual budget and plan, while fixing their attention upon every detail on a daily basis. Putting too much emphasis on the big picture, or on its details alone, would have disastrous effects. You may never complete the picture (being stuck on this or that detail), or you may never paint at all (if all you are envisioning is the final result). Similarly, if your focus is limited to yourself, you may never be well adjusted to your community; if you focus too much on contributing to the well-being of the community, it might exact a heavy toll on your personal life. What I'm saying here summarizes what some of us feel when we devote all our energies to a college or business organization and remain frustrated because we are not appreciated enough or compensated well enough for our efforts. Those of us who limit our horizons to our own lives feel like recluses who are alienated from their community. In these respects, then, education for me is personal and cultural, encompassing institutions of learning as well as other people and documents that tell us stories from which we can learn. This book, therefore, falls into the second category of learning, whether or not it's used in academic settings.

Postmodernism Revisited

Postmodernism, according to what I have just said, could be a fruitful way of thinking that inspires new ideas and that reexamines alleged dead ends. In my version of postmodernism there is room to bring back ancient medical remedies as well as ideas about learning and friendship. The scientific dominance that characterizes the Enlightenment and modernism overshadows past traditions and customs, because they were labeled mystical or irrational, filled with superstition or old wives' tales. Describing an idea in particular ways (homeopathic medicine as inexplicable) has a normative component (it's not worth learning about), and this leads people to ignore potentially useful and informative data. At times our descriptions include a prescriptive element, in the sense that they discourage or encourage further consideration. Some of these tales have a moral; they are suggestive and alert us to basic principles we might otherwise overlook. Therefore, incorporating these ideas and practices can be a safeguard against the disastrous effects of the advent of modernism.

For example, the biblical injunction regarding leaving your land fallow every seventh year is not as silly as it may sound in the age of efficiency and productivity. Obviously agribusiness wants to maximize the production of every acre owned, and there is a tendency to overuse it. But overuse or overproduction can be counterproductive. The nutrients in the soil need to be replenished, because otherwise the fertility of the soil would decrease over time. Growing different crops every year on the same plot of land is an ancient practice of rotation the value of which is being recognized today. Similarly, grandmother's chicken soup does help treat a cold, even though there are no precise scientific justifications for how and why this is the case. "Feed a cold and starve a fever" is a saying the value of which remains intact because of the experiences of many generations. How you implement these insights, how you explain them to yourself and your family may change over time, but they still hold true for most people most of the time around the globe. Though not an expert in these matters, I would venture to say that every culture has its pearls of wisdom handed down over the ages with a great deal of success.

The insights of our elderly and sages are not limited to the medical arena but also pertain to our social and political interactions. We should heed the warnings of those who claim that you shouldn't trust anyone who doesn't have any enemies. That could mean that when one expresses strong opinions or shows courage in undertaking a particular course of action, there are those who will disagree with or be upset by whatever decision was made. These are enemies of sorts (perhaps less personal and more ideological), and as such they are living testimonies to someone else's decision-making processes, or what might be called leadership. Likewise, there are those who claim that you shouldn't trust anyone who clamors for a leadership position but prefer the one who shies away from it. Elect and choose a reluctant leader rather than someone eager whose ambition will overshadow or color the decisions and choices he or she will make. This reminds me of the biblical story of God's choice of Moses to lead the Israelites. Moses was a reluctant leader, professing his inadequacy because of his stutter, for example. Oh, Aaron will be your mouthpiece, was God's response. Don't worry about your own shortcomings, because once you are cognizant of them, they can be corrected, complemented, overcome. Moses, as we know, turned out to be a great leader who had the charisma and personal authority not only to bring the Israelites together but also to leave behind their enslaved state of mind.

Back to postmodernism. This way of thinking also promises to break down hierarchies of power relations among the brokers of knowledge. Dissolving the exclusive appeal of Truth that lay confidently with the sciences and illustrating how alternative ways of seeing the world lead to different pictures and applications of our knowledge, postmodernism was instrumental in leveling the playing field for any new claimant on the scene. Just as the scientific revolutions of the sixteenth and seventeenth centuries took away the power of the church and monarchies so that every claim had to be empirically and rationally defended, so did postmodernism challenge the overwhelming power of reason as the final arbiter between competing worldviews. Not that the church lost all of its power or that reason has become obsolete. Rather, divine revelation and the foundation of logic must wrestle with opposing views and claims as equals in the arena of knowledge politics.

Incidentally, the developments in scientific thought in the past century were meant to have a similar effect on opening the scientific playing field for newcomers. If the truth can be examined by anyone, and if anyone can propose a hypothesis to be open to critical analysis by the establishment, then indeed there is a leveling of the playing field. This was what the Vienna Circle looked for, and this is what Popper and his disciples advocated. The notion of falsification was premised on the idea that any theory, hypothesis, or model could be in principle refuted by anyone, anytime, anywhere. Whether in fact such refutation would be forthcoming in practice is another story. Kuhn reminded us that science is really run by a scientific community, so there are no disembodied theoretical constructs. The scientific community, like any other community, has social, political, and economic components. These components are open to manipulation as well as to power plays and intimidation. As such, they are less egalitarian than hoped for and allow less freedom of challenging expression.

Postmodern thinking tried to remedy these defects later in the previous century. It wasn't a basic disagreement with the ideals of technoscientific research but a supplement to these ideals. Perhaps it was a proposal to improve the methods (and ideally the results) by which we challenge the establishment, the canon, the authority of institutional thinking. When postmodernists welcomed all critical versions from any school of thought, from neo-Marxists (and post-Fordists) to feminism (and postcolonialist), they acknowledged the different quarters from which a critique could be launched. It wasn't about legitimating this or that specific critique but about legitimating any attempt to be critical at all. Some criticisms are less useful than others, some less justified (on some level of justification) than others. But at the end of the day, the very fact that postmodernism opened the door to criticism, to viewing the world differently than before, was in itself a great accomplishment. Moreover, this opening of the door to criticism is also a key ingredient in having a sense of liberty to question and evaluate whatever is presented to us. Besides, if you are invited to put forward your dissatisfaction with the current situation and propose your own view, something radical happens. The burden has shifted to you! You can't simply revert to saying that something or someone is wrong; you have to propose what you think is right.

This is true on a grand theoretical scale as well as on a very practical and personal scale. It's one thing to criticize your neighbor or friend, your coworker or boss, and quite another to explain why you think your criticism is warranted and what should be done differently. It's one thing to complain about the political system, quite another to run for office. One of my old teachers once admonished me that if I was unhappy with the chair of the department or the dean of the college, I should be willing to take their place. "But that's not my point," I said, "I don't want to be an administrator." "But if you complain about their procedures and rules, then how will you change them?" he asked. "Presumably you are not complaining for the sake of it or to cause them grief, right? Presumably you have a better idea of how things should be run at the university! If this is the case, then you have to propose an alternative and be willing to carry it out to fruition, prove that your way is the better of the two." This is also where responsibility comes into play in our own lives.

At the end of day, though, as we have seen time and again, it's not quite clear what is the best course of action. I remember the days when personal computers were introduced to corporate America as the most efficient and user-friendly tools of a new era. I was working at the time for a toy manufacturer, and the implementation was more difficult than anticipated. Of course everyone was relieved about not having to deal with the central computer department anymore, having direct access to inventory data or customer records (of ordering and payment). But this also meant a new burden and perhaps a personal toll on those now using computer technology. What if computers reduce the workforce and some are laid off? What does this mean in regard to the personal responsibility (in committing errors) of those using the computers at their desks? Is everyone now more accountable than ever before, because they can't claim that the computer department screwed things up? The psychological and social constraints and demands introduced by computer technology weren't calculated ahead of time by their promoters. Perhaps these promoters couldn't foresee what was coming; perhaps they didn't care (because they were merely concerned with the technical issues and not the social dynamics of the workplace). Either way, it is an example of how a technoscientific feat changes a cultural landscape (and mind-set) and not only the instruments we use. It's not only what tools we use but how we use them that transform a culture and its inhabitants.

I present postmodernism not as a fashionable mode of thinking of the late twentieth century but as an important lesson of how to contest the supremacy of technoscience, or for that matter any single vision that pretends to be supreme. I present technoscience not as a formidable reality of the twentieth century but as a component of contemporary culture that has to be harnessed for goals other than its own self-perpetuation in academic circles. I bring both to mind, then, in order to figure out what to do about their lingering influence and the advantages we can still gain from keeping postmodernism and technoscience around and in order to make use of them personally in our daily lives.

I would argue that technoscience has taught us to take political and moral action when confronting new technologies and great discoveries of science. We shouldn't remain complacent when the Human Genome Project or stem-cell research is about to unravel the mysteries of the building blocks of human existence. Rather, we should worry how these discoveries will affect our ability to be insured. Will insurance companies demand to know our genetic composition in order to outline and predict our propensity to be prone to contract certain diseases? Will some of us be refused treatment because it is deemed futile according to some medical model that considers our disease terminal? Will our lives be valued differently from those of healthy people, and will, therefore, lack of treatment hasten our demise? These aren't rhetorical questions but real questions that should haunt us today. These are questions that cannot be answered only within the context of medicine and health care but that need taking into consideration political ideals and realities, cultural beliefs and practices. Physicians and hospitals do what is expected of them by society at large; they don't make up their own minds in a vacuum.

Likewise, insurance companies respond to the pressures of the marketplace and their shareholders. When shareholders, not limited anymore to wealthy individuals but including pension funds of labor and teachers' unions, are only concerned with returns on their investments and not with the practices of the companies that they own, we are in trouble. When the horizon of perception is narrowed to quarterly earning reports and not long-term investments and the transformation of the marketplace, then we cannot hope that corporate leaders will behave responsibly. So, insurance companies find themselves less concerned with reimbursements for medical care than with finding reasons not to reimburse and thus to increase their bottom-line profits. Perhaps what gets lost in the process is the very reason insurance companies came into existence to begin with, namely, to collect premiums so that eventual expenses would be fully paid. They were supposed to be our financial custodians and guarantors in case a calamity happened to us. They were never thought of as investment tools or profit makers.

The Ethical Stance

The main reason for bringing up some of the examples mentioned above is to remind ourselves how life can be organized and structured in the pursuit of lofty goals and worthwhile ideals. What I'm talking about are also simple ways, and steps we can take, to recall the original intention behind setting up communities and companies, institutions and political structures. Our congressional representatives are supposed to represent their constituents' concerns and needs, not their own. Teachers are supposed to center their efforts on the most effective learning methods and tools for their students. In all of these—and many other—cases, the positions of power in which some us find ourselves should be secondary and not primary on our mind. We shouldn't think how can I benefit from being a leader but rather how can I better serve others, be available to help others. Humility and dignity, honesty and integrity go a long way toward being the kind of ethical leader I have in mind. Incidentally, by leadership I don't mean only those in highly visible positions of power. I also mean this to apply to those leading their families or coworkers, those involved in educational and religious organizations.

How can we accomplish these ideals? How can we implement them in our daily existence? Here is a framework that might help. First, remember that a universal appeal to any standard (however worthwhile) is problematic in the complex culture in which we live. The demographic make up of our environment has changed over time, and the traditions of our country don't hold sway with those who immigrated from other cultures. This means that compromises are in order and that it's necessary to contextualize every judgment. What applies in one case may not apply in another. Standards change over time, such as the disciplining of children or students or the treatment of women and people of color. What worked well two hundred years ago will not work well today. What works well today in the way we organize and regulate our community will become obsolete in one or two generations. So, the suggestion here is to be open to changes, welcome them, and bring them forth. Contextualization is difficult, because it requires the reflective attitude of a judge who knows the law but needs to apply it each time under different circumstances. Doing otherwise would be unfair, unjust.

Second, the fact that there are multiple stories to tell and multiple standards of ethics to apply means, in the French philosopher Jean-Francois Lyotard's sense (1984, 62), being responsible for one's choices and judgments. It means being responsible for what you say and for the reasons that justify (at least in your mind) saying it. You cannot simply propose ideas or procedures—regarding dress code or voting rules, to name two areas of potential conflict—without being accountable to your proposals. This is different from those (like Feyerabend 1979) who claim that "anything goes," because not any proposal is as valid, good, or practical as any other. Even within a particular context, some proposals make better sense than others, and some can be rationally explained and justified in a manner that would make sense to the majority of the people in a certain community.

Third, under the conditions of cultural ambiguity, the personal investment and commitment of every individual means much more than in ordered societies where all you need to do is follow the rules. When freedom is granted and opportunities for action are less restricted than ever before, there is even more opportunity to be passionate about something (housing for the homeless, care for the elderly, a pollution-free environment). Your passion may wane over time or be transformed from one area to another. But still, it will always be a way to fully insert yourself into the community in which you live. This is what Simone de Beauvoir had in mind after World War II when she spoke of the "ethics of ambiguity" (1991).

As I said earlier, the message of existentialists, like Camus, was not despair and suicide but rather hope and commitment. Likewise, the message of the postmodernists is not silence in the face of holocausts that can be explained within their own contexts by cruel and racist leaders but rather insistence on the declaration of moral convictions and humanitarian acts. Introspection is crucial for the beginning of any consideration of alternatives, so that a decision can be reached that is both reasonable and compassionate. A decision based on introspection and adjudication of competing claims leads then to an action that is defensible beyond the narrow confines of its domain (such as medical expense and prognosis of treatment). You may ask, defensible in what terms?

I would say that the very notion that something is defensible is itself rooted in an Enlightenment belief (or ideal) of how we communicate rationally with each other as equals, how we support our claims, and how we expect others to listen and react to what we say and do. Kant's moral notion of a universal maxim you can generalize from when you do anything rings true in this context of discussion (1981). It's equivalent to the Golden Rule, which states that you are acting morally only when you can allow anyone else in your situation to do the same as you do. But the generalization is not as simple as Kant envisioned it some two hundred years ago, because we can assume neither a universal logic nor a universal language with which to internationally communicate with our neighbors, as we see daily in the deliberations of the United Nations (itself an Enlightenment ideal). We have added layers of complexity over the years, and by this century we find ourselves dwarfed by what we must know to even begin a conversation. The task of communicating seems too daunting, because of cultural differences, so at times we remain silent. We even make light of questions of life and death because we don't feel up to the task, we don't know enough. Leave it to the experts to sort out what we should do, we say to ourselves and to our neighbors. But the experts don't know for sure either. Even if they do, do we really wish to abdicate our power to answer or shirk our responsibility to answer?

Perhaps what we must talk about at the close of this book is personal responsibility for our own actions and the actions of our community. Neighborhoods and communities are themselves realities that are socially constructed and culturally contested. For purposes of pollution prevention, a "neighborhood" can cross county, state, and national borders. As for neighborhoods, they can be geographically circumscribed or virtually woven through Internet connections throughout the wired world. As for the self, as sociologists and psychologists will attest, this too is a contested territory, depending on one's views, experiences, and prejudices. How much of ourselves do we know and secure from external influences? How independent and isolated is the individual in a culture bent on mass communication and the obliteration of the dividing lines between the public and personal spheres? What about the notion of responsibility? Do we mean by it taking responsibility for our thought and action, or for our fellow citizens? So, even the question I ask about personal responsibility is open ended and somewhat ambiguous. Where do you go from here? What are you going to do? When will you take charge of your own fate?

~

Works Cited

Agassi, Joseph. 1971. *Faraday as a Natural Philosopher*. Chicago and London: University of Chicago Press.

Arendt, Hannah. 1958. *The Human Condition*. Chicago and London: University of Chicago Press.

Aristotle. 1985. *Nicomachean Ethics*. Trans. Terence Irwin. Indianapolis: Hackett.

Ayer, A. J., ed. 1959. *Logical Positivism*. New York: Free Press.

Bacon, Francis. 1985. *The New Organon* [1620]. New York: Macmillan.

Bauman, Zygmunt. 1991. *Modernity and the Holocaust* [1989]. Ithaca, N.Y.: Cornell University Press.

Berlin, Isaiah. 1969. "Two Concepts of Liberty." In *Four Essays on Liberty*. Oxford and New York: Oxford University Press.

de Beauvoir, Simone. 1991. *The Ethics of Ambiguity* [1948]. Trans. B. Frenchman. New York: Citadel.

Brillat-Savarin, Jean-Anthelme. 1970. *The Physiology of Taste* [1825]. Trans. A. Drayton. New York: Penguin Books.

Camus, Albert. 1991. *The Myth of Sisyphus and Other Essays* [1942]. Trans. Justin O'Brien. New York: Vintage.

Dewey, John. 1960. *The Quest for Certainty: A Study of the Relation of Knowledge and Action* [1929]. New York: Capricon Books.

Epictetus. 1983. *The Handbook (The Encheiridion)*. Trans. Nicolas White. Indianapolis: Hackett.

Feyerabend, Paul. 1975. *Against Method: Outline of an Anarchistic Theory of Knowledge*. London: Verso.

Foucault, Michel. 1970. *The Order of Things: An Archeology of the Human Sciences* [1966]. New York: Vintage Books.

———. 1979. *Discipline & Punish: The Birth of the Prison* [1975]. Trans. A. Sheridan. New York: Vintage Books.

Frankl, Viktor E. 1959. *Man's Search for Meaning* [1946]. New York: Washington Square.

Freud, Sigmund. 1989. *Civilization and Its Discontent* [1930]. Trans. James Strachey. New York: Norton.

Gleick, James. 1987. *Chaos: Making a New Science*. New York: Penguin Books.

Hawking, Stephen. 1988. *A Brief History of Time*. New York: Bantam Books.

Hobbes, Thomas. 1968. *Leviathan* [1651]. Ed. C. B. Macpherson. New York: Penguin Books.

Hoesterey, Ingeborg, ed. 1991. *Zeitgeist in Babel: The Postmodernist Controversy*. Bloomington: Indiana University Press.

Holton, Gerald. 1996. *Einstein, History, and Other Passions* [1995]. Reading, Mass.: Addison-Wesley.

James, William. 1955. *Pragmatism* [1907]. Cleveland and New York: Meridian Books.

Kafka, Franz. 1969. *The Trial* [1937]. New York: Vintage.

Kant, Immanuel. 1970. An Answer to the Question: "What Is Enlightenment?" *Kant's Political Writings*. Trans. H. B. Nisbet. Cambridge: Cambridge University Press.

———. 1981. *Grounding for the Metaphysics of Morals* [1785]. Indianapolis: Hackett.

Koyre, Alexander. 1968. *Newtonian Studies*. Chicago: University of Chicago Press.

Kuhn, Thomas. 1970. *The Structure of Scientific Revolutions* [1962]. Chicago: University of Chicago Press.

Lasch, Christopher. 1991. *The True and Only Heaven: Progress and Its Critics*. New York and London: Norton.

Lyotard, Jean-Francois. 1984. *The Postmodern Condition: A Report on Knowledge* [1979]. Trans. G. Bennington and B. Massumi. Minneapolis: University of Minnesota Press.

Marx, Karl. 1977. *Capital* (Vol. I) [1867]. Trans. Ben Fowkes. New York: Vintage.

Nietzsche, Friedrich. 1967. *The Will to Power*. Trans. W. Kaufmann and R. J. Hollingdale. New York: Vintage.

Oppenheimer, Robert J. 1955. *The Open Mind*. New York: Simon and Schuster.

Orwell, George. 1961. *1984* [1949]. New York: New American Library.

Plato. 1974. *Republic*. Trans. G. M. A. Grube. Indianapolis: Hackett.

Pool, Robert. 1997. *Beyond Engineering: How Society Shapes Technology*. New York and Oxford: Oxford University Press.

Popper, Karl R. 1959. *The Logic of Scientific Discovery* [1935]. New York: Harper and Row.

———. 1962. *The Open Society and Its Enemies*. Princeton, N.J.: Princeton University Press.

Postman, Neil. 1992. *Technopoly: The Surrender of Culture to Technology*. New York: Vintage Books.

Prigogine, Ilya, and Isabelle Stengers. 1984. *Order out of Chaos: Man's New Dialogue with Nature*. New York: Bantam Books.

Rousseau, Jean Jacques. 1964. "Discourse on the Sciences and Arts." *The First and Second Discourses*. Ed. and trans. Masters. New York: St. Martin's Press.

———. 1964. *On The Social Contract*. Ed. and trans. Masters. New York: St. Martin's.

Sartre, Jean-Paul. 1976. No Exit *and Three Other Plays* [1945]. New York: Vintage.

Sassower, Raphael, and Charla P. Ogaz. 1991. "Philosophical Hierarchies and Lyotard's Dichotomies." *Philosophy Today* 2:153–60.

———. 1993. *Knowledge without Expertise: On the Status of Scientists*. Albany: SUNY Press.

———. 1995. *Cultural Collisions: Postmodern Technoscience*. New York: Routledge.

———. 1997. *Technoscientific Angst: Ethics and Responsibility*. Minneapolis: University of Minnesota Press.

———. 2000. *A Sanctuary of Their Own: Intellectual Refugees in the Academy*. Lanham, Md., and Boulder, Colo.: Rowman & Littlefield.

Schumacher, E. F. 1973. *Small Is Beautiful: Economics as if People Mattered*. New York: Harper and Row.

———. 1977. *A Guide for the Perplexed*. New York: Harper and Row.

Shelley, Mary. 1967. *Frankenstein* [1818]. New York: Bantam Books.

Snow, C. P. 1964. *The Two Cultures and a Second Look*. Cambridge: Cambridge University Press.

Stephan, Paula, and Sharon Levin. 1992. *Striking the Mother Lode in Science: The Importance of Age, Place, and Time*. New York and Oxford: Oxford University Press.

Veblen, Thorstein. 1899. *The Theory of the Leisure Class*. New York: Macmillan.

Weber, Max. 1968. *Economy and Society: An Outline of Interpretive Sociology*. Berkeley: University of California Press.

Winner, Langdon. 1977. *Autonomous Technology: Technics-Out-of-Control as a Theme in Political Thought*. Cambridge, Mass.: MIT Press.

Wittgenstein, Ludwig. 1959. *Tractatus Logico-Philosophicus*. Trans. C. K. Ogden. London: Routledge and Kegan Paul.

~

Index

~

About the Author

Raphael Sassower is Professor of Philosophy at the University of Colorado, Colorado Springs. Though his area of prime interest is postmodern technoscience, he has written on other topics as well. Most recently, his publications include: *Cultural Collisions* (1995), *Technoscientific Angst* (1997), *The Golden Avant-Garde: Idolatry, Commercialism, and Art* (2000), and *A Sanctuary of Their Own: Intellectual Refugees in the Academy* (2000).